Springer Series on Environmental Management

Robert S. DeSanto, Series Editor

Springer Series on Enviromental Management
Robert S. DeSanto, Series Editor

Robert S. Dorney

The Professional Practice of Environmental Management

Edited by Lindsay C. Dorney

With 23 Figures

Springer-Verlag
New York Berlin Heidelberg
London Paris Tokyo

Robert S. Dorney‡

Lindsay C. Dorney
University of Waterloo
Waterloo, Ontario
Canada N2L 3G1

Library of Congress Cataloging in Publication Data
Dorney, Robert S., d. 1987.
    The professional practice of environmental management/Robert S.
Dorney : Lindsay C. Dorney, editor.
    p. cm.—(Springer series on environmental management)
    Bibliography: p.
    ISBN 0-387-96907-1 (alk. paper)
    1. Municipal engineering. 2. Environmental protection.
I. Dorney, Lindsay. II. Title. III. Series.
TD149.D67 1989
628—dc19   88-38910

Printed on acid-free paper.

Typeset by TCSystems, Inc., Shippensburg, Pennsylvania.
Printed and bound by R.R. Donnelley & Sons, Harrisonburg, Virginia.
Printed in the United States of America.

9 8 7 6 5 4 3 2 1

ISBN 0-387-96907-1 Springer-Verlag New York Berlin Heidelberg
ISBN 3-540-96907-1 Springer-Verlag Berlin Heidelberg New York

*To my wife Lindsay,
for showing me the symbolic landscape*

*We . . . need to know more than ever before—and indeed more than we
know—to imagine who and what we are with respect to earth and sky. I
am talking about an act of imagination essentially, and the concept of a
. . . land ethic.*

— N. Scott Momaday, "The Names: A Memoir"

Series Preface

This series is dedicated to serving the growing community of scholars and practitioners concerned with the principles and applications of environmental management. Each volume is a thorough treatment of a specific topic of importance for proper management practices. A fundamental objective of these books is to help the reader discern and implement man's stewardship of our environment and the world's renewable resources. For we must strive to understand the relationship between man and nature, act to bring harmony to it, and nurture an environment that is both stable and productive.

These objectives have often eluded us because the pursuit of other individual and societal goals has diverted us from a course of living in balance with the environment. At times, therefore, the environmental manager may have to exert restrictive control, which is usually best applied to man, not nature. Attempts to alter or harness nature have often failed or backfired, as exemplified by the results of imprudent use of herbicides, fertilizers, water, and other agents.

Each book in this series will shed light on the fundamental and applied aspects of environmental management. It is hoped that each will help solve a practical and serious environmental problem.

Robert S. DeSanto
East Lyme, Connecticut

Preface

"Without landscape, we have no identity," Robert Dorney said in an interview not long before he died. The environmental events of 1987 and 1988 have forced us to look at our identity as consumers and polluters and to recognize that no landscape can absorb continued abuse. In a plot worthy of a novel by Joseph Conrad, a New York City garbage barge searched the east coast of the United States for a site to disgorge its rotting and malodorous freight; ships from Europe looked to west African countries to welcome their similar cargo. Medical waste washed up on the Atlantic beaches—an ugly and dangerous reality intruding into a landscape sacred for recreation and sun worship. A hole was discovered in the Antarctic ozone layer, dry forests were incinerated, and floods swept hundreds of people away. Nature seems to be flinging back the abuse; identity is problematic.

Unlike these events which make headlines and prime time television news, quietly land uses change and development moves ahead. Government agencies approve development plans; hearing boards review objections from landowners, neighborhood citizen groups and nature organizations. What will be the impact of development on the environment is the usual focus of the objections; each complainant has its scenario of degradation. How is the potential impact determined? What is the process by which the kind and degree of environmental impact are assessed? This book addresses those issues and does it in a manner and language which scientists, designers, the legal profession, regulatory agencies, and the public understand. The key role in that process is that of the professional environmental manager.

However, Robert Dorney intended that this book should go beyond that important and basic point. With this book he wanted to change the structure of environmental discourse, if not to create a new one, responsive to environmental needs and goals of a community. The traditional structure of the discourse is one of confrontation and advocacy, dominated by the legal profession. One party pleads a case or issue before an adjudicating body; the other parties counter that position. A ruling is made as to the rightness or wrongness of the case. Dissatisfied parties launch an appeal, and the case moves up the hierarchy of adjudicating bodies at considerable expense to all involved, including the taxpayer. Instead, in the author's vision, what is required in development issues and proposals is a way to generate a range of environmental alternatives, from which one or two would be selected. Which one or two are the most appropriate would be determined by the environmental and social goals evolved from the political process and long-range planning. Such is the *ideal*, in which citizens, scientists and social scientists, planners and designers, lawyers and politicians must engage. This book, describing and analyzing a process by which environmental managers can determine the scale and data (physical, biological, and socio-cultural) necessary, interpret the data and develop several environmental scenarios for consideration, is the genesis of this new discourse. Every participant in the discourse knows and understands what an environmental manager contributes to the planning and development process. This new discourse is built upon an open systems approach, which was the configuration of the author's mind from the 1961 international award-winning study of ruffed grouse to his last publication on environmental education.

What Robert Dorney has written will seem obvious and simple, like a cookbook detailing fundamental techniques. Twenty years of experience in academia as a professor in the School of Urban and Regional Planning at the University of Waterloo and 18 years as an international environmental scientist, consulting ecologist, founder and chair of the board of Ecoplans Ltd, and expert legal witness have been distilled in this book. As well, on his several sabbatical leaves to Turkey, Iran, and Costa Rica, he was to test his ideas concerning landscape evolution in diverse topographies and climates and in differing cultures.

Ecoplans Ltd was the result of Dorney's commitment to applying environmental science to the planning and development process, to create a better landscape in which to live, and to renew those which have become derelict. In his writing and lectures, he wove together theory, methodology, and application: For his students he wished to make some "wise cracks in the ivory tower." Shaking the tower is strenuous. The lack of grants from the agencies charged with research funding forced him into establishing Ecoplans: The science agencies rejected his research as too socially oriented; the social science agencies thought his research too scientific. Through projects contracted to Ecoplans, he was able to explore and demonstrate how environmental science can be the basis of rational, less costly, and aesthetically pleasing development. Over the years these projects, nearly a hundred in number, ranged in type from urban, institutional, and industrial, to engineering and servicing, to landscape architecture and landscape renovation, agricultural studies, and corridor studies, to recreation, parks, and open space studies, and to policy development at provincial and federal levels.

Sadly, he did not write this preface; it would have been a different one. My contact with Ecoplans has been a close, if not a domestic, one. He did not mention that characteristically consulting firms begin small, literally in the home (in fact in the bedroom), then grow until they are moved out of the household into official office space. During the early years of Ecoplans I did much of the historical research and interviews as well as editing studies and reports. The role of the communications specialist was a key one he came to appreciate. It is now ironic that I have returned to that role in order to work through his manuscript.

The acknowledgments which Robert Dorney listed are numerous and reflect both the diversity of his professional colleagues and friends involved in environmental projects and his generosity: ". . . to my colleagues in Ecoplans Ltd past and present, who added their talent to exploring the boundaries of environmental management: Mike Allen, Sid Amster, Douglas Hoffman, Mike Hoffman, Cam Kitchen, Susanne Kitchen, Grace Lawson, John MacArthur, Hugh MacCrimmon, Rob MacGregor, Vince Moore, George Mulamoottil, George Rich, and Ray Smith. Of those colleagues in academia, government and industry who contributed ideas and the opportunity to implement our ideas, the following people were particularly helpful: Jim Bater, John Bousfield, Doug Calder, Bill Coates, Derek Coleman, George Cornwell, Merve Couse, Cheryl and John Dorney, Mike Fearnall, Len Gertler, Steve Glogowski, Eric Higgs, John Leicester, Ian MacNaughton, Jim Murray, Bob Narin, Doug Nelson, Gordon Nelson, Tom Owen, John Page, Mel Plewes, Henry Regier, John Seldon, John Sutherns, Forest Stearns, Bill Thomson, and Keith Wilde."

Many of those are friends to whom I extend my deepest gratitude for their patience, guidance, thoughtful critique of my efforts, and belief that this project could be completed—George Rich, Doug Hoffman, and Cam Kitchen; to Dr. Julian Dunster and Dr. Eric Higgs, Robert Dorney's last graduate students, for their enthusiasm and deep appreciation of their advisor's work and passion for the environment and for their friendship to me; and to Grady Clay for his "Foreword" which places Robert Dorney's innovative work in the context of the environmental movement, and to Lawrence Haworth for the "Afterword," which examines the idea of "creative disequilibrium."

Lindsay C. Dorney
University of St. Jerome's College
University of Waterloo
Ontario, Canada

Foreword

Jostling for position, during the 1970s, among the newly aroused-and-aware Earth Day groups, the ecofreaks, and assorted old-line conservationists, a few voices began to be heard above the din: that of Ian McHarg, his eloquent Scots brogue arousing students into frenzies of attention at the University of Pennsylvania; Paul Ehrlich, letting population bombs loose at seminars; Garrett Hardin, making uncommon commotion over the loss of common rights; and others who seemed to be specializing in manipulating the media as much as massaging their own message of coming environmental doom.

Suddenly, having been exposed to the first mutterings of the distant pack, and then the louder propaganda of their successors, I began to be aware of other voices in other rooms, especially Canadian. The name of Robert S. Dorney kept reappearing among intriguing footnotes and conference minutes that flowed into my office as the editor of a magazine then tracking the new wave. We had done our own bit for the cause, devoting an entire issue of *Landscape Architecture* magazine to "Ecology as a Basis for 20th Century Planning and Design." That was in January 1967. I had already been alerted to the fact that Canadians were marching to their own drummer (which beat no calls to arms for the Vietnam War, which was wrenching apart the not-so-United States).

Canada was, and remains, a vital balance wheel to its rampageous southern neighbor. The Canadian impetus to restrain the rampant forces of damn-all-development differed in the 1970s from what was happening south of the border. Canadian land in the path of urban sprawl and speculation was under pressures similar to those in the United States. But there were significant differences: Urbanization was not perceived to be the universal threat to resources as was being accepted in older-settled (New England) and newer (California) regions of the United States. Canada's economy was, after all, heavily resource-based, its timber and mineral companies holding great political power. Urbanization was not the national preoccupation it was in many parts of the United States, for Canada's urbanization corridor was an attenuated one, stretching for hundreds of miles along the Great Lakes, the St. Lawrence River, and the Pacific Coast. Outside all this ferment, the back country loomed huge, richly endowed with forests and minerals, traditionally inaccessible but now coming under attack from every form of corporate raiding. Canada's great West stood and felt isolated and resentful. The Arctic was coming under incredible point-source pressures from oil drillers, ice breakers, motorized hunters, and other exploiters of native territories and rights.

As a "transplanted" American, Dorney had to adjust to Canada's political and cultural dynamics, as well as its forms of urban development, which are different sorts from those of Europe or the United States. Furthermore, he had to overcome the usual reaction to so-called experts from the United States who were greeted coldly if at all by nationally conscious Canadians who still (until 1988) would search U.S. automobiles at the border for architectural plans and drawings illegally smuggled to compete with home-grown species.

In this milieu, the time gradually ripened for locally trained experts, flush with regional expertise in the realities of landscape change. As distinct from their U.S. counterparts, Canadian government planners and even academic consultants could benefit from a long tradition of British-trained and usually incorruptible civil servants. Spokesmen for local, provincial, or national governments were expected, with good reason, to offer sane, sound, hard-headed, and locally based (even if locally unpalatable) advice.

In such a milieu, the voice and writings of Robert Dorney soon began to acquire impact. He spoke from a solid base of teaching at the University of Waterloo, backed by private practice as a consultant, and as owner of a landscape nursery—a rare combination of expertise. I first sighted him at an international conference of landscape architectural students in the early 1970s. The Dorney name on the program aroused my anticipation. The Dorney person loomed impressive at a distance. Then, suddenly, as a note-taking editor in the audience, I found myself listening with every fiber as this tall, broad-backed man shared his no-nonsense observations compellingly and in pungent, often witty, and first-hand detail. So brisk was his pace, so pell-mell his citations, that the student audience that afternoon was stunned, and questions were few. I had come prepared for the impetus, however, and, with tape recorder and other aids, ran through a detailed interview with Dorney. That was the beginning of a long friendship, nurtured inside and outside universities and formal conferences.

On another such occasion, in a totally darkened room, I listened enthralled as Dorney recounted bird counts and detritus samplings along the metropolitan fringes of Toronto. From all this emerged the conclusion that abandoned subdivisions and shopping center rooftops were among the most variegated and well-populated bird habitats of Ontario. Here was ill-tasting medicine, indeed. For among traditional conservationists there were few objects of scorn more useful for propaganda purposes than raw-and-jammed shopping centers and the raw-and-unkempt reality of abandoned housing tracts. Yet the facts were as stated: so-called "marginal" sites can serve many masters and complex purposes when examined by the ecologist's depth perception and systems view.

Dorney further described how his students had identified a "hawk ring" 12 miles out of Toronto. This turned out to be an exurban belt of abandoned farmland—heavily taxed and with too many scattered homes for efficient farming, but not yet cut up into housing tracts. There it stood, land going to seed, growing up in brush—a haven for field mice, snakes, rabbits, and other wildlife. Overhead, sharp-eyed hawks hovered in sufficient numbers to be identified by sharp-eyed observers. The room was quite dark, which made note taking impracticable. My journal merely and lamely records "Superb! paper."

Michel Foucault would have found much to admire, as I did, in the intensity of gaze that Dorney focused upon throwaway sites, oddments of incidence, collections of happenstance, the ephemeral amid the long-standing. This great curiosity on Dorney's part was backed up by the full apparatus of public teaching, research, and private practice. Truth thus apprehended from several directions becomes more complex; it can be enlivened by the touch of many disciplines. Mixed with the Dorney wit, it could be compelling.

Not the least of advantages that go with this varied life was the self-discipline acquired by testimony before tribunals. As Dorney's writings on the following pages make clear, it is one thing to write for scholars, but another thing to write knowing that one's work and person may well end up in court. Ecology became a hot property in the 1970s, and ecological reports were required to support or to challenge plans for new projects—dams, factories, pipelines, subdivisions. There always loomed around the corner a possible lawsuit or hearing by an administrative tribunal and cross-examination to challenge not only one's findings but also one's right to be known as an expert. Such fine tuning on the witness stand seemed to bring out strengths that Dorney reapplied to his research. And I suspect it earned him the envy as well as the respect of many colleagues from the less litigious realms of academia.

All these encounters were high points in the transition of ecology from its in-group beginnings as a biologists' monopoly to become a mixed enterprise of great public value. As an editor in far-off Kentucky, I watched with admiration Dorney's expanded career. He was a trail blazer, willing to put his money where his brain was, investing in offices in a suburban industrial district, using his sideline, the plant-supply business, for test runs; and teaching, consulting, team forming, and multi-laning.

Unlike too many of his U.S. counterparts, Dorney brought to his teaching (and brings to this book) an international viewpoint. To sit among Canadian planning and designing teams is to hear citations from an international array of documents and anecdotes from a constellation of jobs and joint ventures. Many of these are based on old Empire or Commonwealth connections. Long before their cousins to the south, Canadians played the game of international consultancy with verve and multilingual skills.

Increasingly Dorney was asked to give papers at international conferences, one of the last in Japan in 1987. He appeared to us observers as one of a still small but expanding band of pioneers. The band included people of many origins (Dorney's doctoral degree, as I recall, was in veterinary science from the University of Wisconsin). Others rose to the top of this new specialty from other sciences and the design arts. But all who approached Dorney's level were multiplex persons, pushing the limits of old job descriptions, sometimes upsetting or rearranging old-line organizations.

Rarely are such persons willing and able to write carefully about the daily necessities of a pioneering form of professional practice. We are all lucky that, before his untimely death at the age of 59, Dorney took time to leave a useful and widening paper trail, to share an insider's knowledge of this-is-how-it-all-worked, and to suggest ways one man's rich experience can energize an expanding profession.

Grady Clay

Grady Clay was editor of the international journal *Landscape Architecture* from 1960 to 1984, and is the author of several books on urban environmental affairs.

Contents

1

Introduction: The New Profession of Environmental Management

The explosion of a worldwide environmental consciousness in the late 1960s and into the 1980s is a social phenomenon of amazing intensity. The near hysterical levels reached by some ecological prophets had strong parallels with certain religious sects that have prophesied the end of the world, sold all their earthly possessions, and then waited for the Day of Judgment. Fortunately, global ecosystems were not ready to quit, as yet, so we must assume they had more basic resilience than has been attributed to them.

Historically, conventional or operational ecological wisdom evolved in regard to land management, which was strongly embedded in folklore and rural tradition in the agrarian societies (Caldwell 1970). However, in the urban and industrial society that emerged over two centuries, there was a cultural lag in ecological wisdom, a lag that was finally perceived in the 1960s and that became epitomized by the term "Spaceship Earth." This new awareness, fueled in part by *The Silent Spring* (1962) by Rachel Carson, space exploration, and radioactive deaths from military use of atomic weapons, exposed raw nerves which when probed caused convulsions and twitchings among politicians, social scientists, and natural scientists.

Not surprisingly, technocrats, especially planners, engineers, and economists who had ignored earlier advice from conservationists, bird watchers, and "little old ladies in tennis shoes" were caught off guard when it was proposed that *Homo sapiens* was moving toward extinction within the next century. However, if this wave of hysteria demonstrated nothing else, it was that glaring gaps in technical information existed, such as in the magnitude of the CO_2 greenhouse effect, danger of genetically altered organisms released into nature, DDT, the SST and ozone debate, and, more recently, the effects of acid rain, polar ozone window, dioxin, and potential for nuclear winter. Sharp differences in interpretation of skimpy data were not unusual. More sophisticated levels of chemical testing to parts per billion identified new potentially toxic compounds requiring additional studies.

The hysteria and data gap also demonstrated that the scientific and engineering specialties were often ill prepared to conceptualize a problem in a holistic way, let alone consider priorities for eventual solution. Disagreement among academic ecologists as to the underlying causes for environmental deterioration, such as the population growth versus technology debate, also emerged. The public distrust of scientists and engineers suggested that scientific information could lead to serious overconfidence and in some cases endanger human life. The Chernobyl and Three Mile Island nuclear accidents and the loss of the space shuttle Challenger epitomized this distrust of and loss of confidence in technocrats.

From an administrative and political viewpoint, it was clear that some reconciliation between the *market point of view,* espousing growth and progress, and the *ecological point of view,* espousing a dynamic equilibrium between man and nature, was imperative (Caldwell 1970). The issue was not, obviously, one or the other, but how to capture ecological wisdom without destroying all the cultural evolution, technology, and urbanization that have been proceeding unabated for over 200 years and that now include the less industrialized nations of the Third World. Rather, the issue is how to avoid plunging the developed world into a new dark age because of technology producing irreversible ecosystem* repercussions, thereby resulting in irreversible social and economic disintegration, or how to allow the poorer countries to raise their standard of living without massive destruction of forests, soil, and fisheries resources and massive pollution. The (Brundtland) World Commission on Environment and Development report in 1987, for both the developed and the less developed countries, demonstrated that good environmental management is achievable without substantial sacrifice in the standard of living.

If the public makes any sense out of the continuing furor, it would be articulated by the cynics as resulting in higher prices for energy and automobile pollution devices, while the optimists would point with pride to the number of flattened tin cans taken to the recycling center or to the number of new energy-efficient homes built, or comment on the number

---

* An *ecosystem* is an open system having transboundary flows of energy and matter. It has in addition an identifiable biophysical boundary as well as internal dissipative structures, geochemical and biological processes, and homeostatic mechanisms.

of new environmental guidelines passed by the Environmental Protection Agency or the city council.

To the natural scientist doing detailed research on bioenergetics of hummingbirds or on population dynamics of the intestinal coccidia of flying squirrels, speaking on the ecological crisis to a service club or a church group can be traumatic to both the ecologist and the public. The experience can indeed cause a few cracks, or wisecracks, in the ivory tower. From the public's point of view, these encounters demonstrate that much of the academic ecology being funded with public money was, if not irrelevant, at least trivial in view of the near-ending or impending crises. Furthermore, it caused many scientists to question their own commitment and relevance in applying for another grant to study nature in vacuo—that is, nature devoid of cultural intrusions.

As academic scientists emerged in increasing numbers over the past 15 years from the narcotic cocoon of the research grant and biology department structure into unemployment or underemployment, they found not only that their students were ready to shed biological "skins" but that employment as professionals was possible. Such a metamorphosis, however, requires an interprofessional dialogue under the emerging concept of ecosystem management: scientists could accept that human beings are part of the ecosystem and that other professions dealing specifically with the human condition were not so villainous after all. In short, recognizing that the world was slightly more complex than merely being divided into two camps: the good guys, that is ecologists, natural resource analysts, and conservationists; and the bad guys, that is, industrialists, developers, and their allied professionals—lawyers, planners, architects, engineers, and economists.

The conceptual hiatus between urban planning and urban ecology is admirably demonstrated by the plethora of books on the subject of urban planning and design (see Bacon 1967). In such basic textbooks, management of the urban natural environment is discussed under the "need" for parks or open space, the advisability of saving a few trees because they can be decorative, or the "need" to restrict development on hazard lands: floodplains, steep slopes, etc. The urban environment is perceived in terms of solely structural elements, such as streets and buildings. In all, there is no awareness that the natural environment is a system and that systems ecology theory, embracing both structure and process, can be extended to the urban physical-natural environment. Conceptualizing the urban or urbanizing environment in a systems context enables planners and engineers to improve substantially their level of professional practice. The quality of the new urban environment could thereby be greatly enhanced by embracing this holistic or systems viewpoint (Dorney and Rich 1976; Dorney 1983; Boyden et al. 1981; Hough 1985; Sprin 1984).

The symposium on *Future Environments of North America* (Darling and Milton 1966) was one of the first interprofessional debates to so clearly reveal the frightening isolation of academic ecology and resources management from the mainstream of society. It did, however, suggest that the gap could be narrowed if not abridged.

Another early bridge so admirably flung across the conceptual chasm between this "new" ecology and the design professions was the work of Ian McHarg (Darling and Milton 1966; McHarg 1969). In addition, for the field of technology and ecology, Barry Commoner (1971) suggested that a complete restructing of the present economic system, utilizing science and its technology, would be needed for the United States. Whether or not these approaches will bridge or widen the chasm between economics, engineering design professionals, and ecology remains to be seen, but I believe it to be, a somewhat sturdy bridge, and individual success stories exist (Dorney et al. 1986).

Management sciences or operations research, another approach to problem solving developed from World War II, applied scientific knowledge to a wide range of problem solving (Papageorgiou 1980). In this sense it is another systems science area similar to environmental management, but it covers a much wider area of applied sciences than just those concerned with the environment, and it evolved much earlier.

Simultaneously, after World War II the physical and biological sciences were evolving technologies that could be applied to measuring or sensing environmental realities. Infrared photography and radar scanning, satellite capability, deep ocean drilling, lenses with high resolution for high-altitude work, sophisticated chemical analyses such as spectrometry, computer-based biophysical models—all developed from post–World War II research and development expenditures. The IBP (International Biological Program) and MAB (Man in the Biosphere) programs brought scientists from many countries into discussion with agricultural, physical, and social scientists at the National Academy level of decision making, and required interscientific disciplinary cooperation. These new analytical and organizational skills, quickly put to use, reinforced the Spaceship Earth concept.

Understandably, the first public response to the environment crisis was to pass laws on pollution restricting the activities of the "bad guys." Others called for a rejection of the growth ethic, blaming it on Judaism, Christianity, and any other ism that was handy. The "conserver society" was viewed as the answer, although high interest rates and high unemployment in the 1980s may have reduced consumption more by necessity than by choice. Consumerism as an allied area of interest similarly threatened the large industrialists with a more restrictive legislative environment where industry was asked to justify not only its advertising and its ethics but its environmental concerns as well.

In the early 1980s the emergence of the Green political movement, or its equivalents in Europe and North America, adds a new dimension to the environmental movement. Since antiwar and anti–nuclear missile sentiments are intertwined with concern for growth and industrialization, it is as yet unclear how ecosystem analysis and management concerns will fare. High current unemployment and the young age of the Green movement suggest the evolution of a postindustrial society may be accelerated by this political synthesis of concerns about pollution, nuclear war, feminism, and the growth ethic. At least in West Germany the Greens will have a chance to demonstrate whether it can offer constructive solutions or simply disruptive ones. Or, put in another perspective, whether "radical ecology" or "deep ecology" will assist in offering viable options such as ecodevelopment and bioregionalism, or simply engage in industrial sabotage and obstruction.

New government and academic structures emerged in both North America and Europe. Environmental studies were born in the universities—not as legitimate, perhaps, as the traditional departments, but nonetheless their presence is felt. Governments consolidated natural resources, water, solid waste, and air management branches into departments of the environment. Suddenly engineers were thrown in with fishery and wildlife biologists; foresters eyeballed health officers and planners—bolstering the sale of tranquilizers if nothing else.

Concurrently, environmental professionals began to develop private consulting practices. Drawn from the social, natural, engineering, design, and geographic sciences, they had some common threads: a systems view, a human ecology view, an environmental ethic, and a willingness to work for private, government, or community groups in a political and legal context. I now believe these professionals should be identified under the generic label *environmental managers*.

To be successful, environmental management requires an institutional, economic, social, technological frame of reference from a systems perspective, as well as from a series of time perspectives (past, present, future), a view shared by Petak (1980, 1981). Environmental management also has to be adaptable to political swings of the electorate, left to right, and to economic swings of inflation and deflation. It requires short-term goals embedded within an overall vision of the achievable future. Its implementation requires a legal and a democratically elected political system able to adjudicate arguments of social justice and technical information which recognizes ecosystems dynamics (Odum 1983) and the adaptive management approach (Holling 1978). Politicians must be alert to issues of environmental quality and support research and development efforts in this regard. Public education is vital.

For autocratic governments—left and right having centralized state planning—the legal and political process is differently regarded: environmental quality concerns are argued almost exclusively in the government sector. Implementation may occur with minimal public information and knowledge. When the government changes, abruptly in many cases, the new central planning groups may sweep away all former environmental controls. This suggests that autocratic governments are more unpredictable in their environmental management, positing that pollution is solely a product of capitalistic economies and practices.

Now that we are well through this first turmoil created by the international environmental hysteria, or realism, of the last 15 to 20 years, a number of generalized professional and societal priorities applicable not only to North America but to many other industrial and industrializing countries can be distinguished. Some of these priorities are already generating demand for environmental managers. Predictions can be made for the industrialized countries of North America, Eurasia, Australia and New Zealand, and for parts of Africa, Latin America, Southeast Asia, and the Middle East that create a demand or justification for more professionally trained environmental managers. A reader can see the humanistic promise offered by this approach, and the cautious optimism that is implied. Unless we identify ecosystem stresses and control at varying scales, we all can become part of the problem. I believe humanistic solutions in an ecodevelopment framework are possible. That is, a human-ecology framework of environmentally sensitive change is achievable. This book will identify how, in part, this can be accomplished.

Justification for This New Profession in Industrialized Countries

For the industrialized and technologically advanced nations, the following 13 considerations should provide the thrust to increase the need for a range of environmental managers oriented toward environmental quality issues.

1. The present professional training of engineers, lawyers, doctors, architects, planners, and some landscape architects continues to be enriched by systems ecology input relating issues of land use, human density and health, industrial activity, and natural resources. To put it another way, the reductionist point of view often used in problem solving by scientists, doctors, veterinarians, and engineers will be balanced in professional training by a holistic point of view put forward in part by new in-house environmental staff. Such a promising beginning, as discussed by Caldwell (1982, p. 110), has already been made in planning and landscape architectural schools in many countries. Environmental engineering and systems design engineering also bridge in part this need to train specialists who are technically competent but who also understand the importance of contextual knowledge.

2. The shortage or virtual absence of base mapping on the qualitative and quantitative distribution of natural and human-made systems

(forests, water bodies, agricultural land, urban open space, etc.) will generate a demand for its creation for environmental planning purposes. Existing mapping usually does not have comparable or suitable scale; often maps are out of date. The ERTS and Landsat satellites, one new tool making such up-to-date land use coverage more feasible, require environmental scientists and geographers of many backgrounds to interpret the data.

3. The need will increase to do intensive and extensive environmental monitoring by both the private, government (regulatory), and labor union sectors. The absence of monitoring during the past 20 to 50 years produced an information vacuum no longer tolerable and too risky to ignore.

4. The need to restore economically derelict lands, such as strip-mined areas, hard-rock mining areas, gravel pits, riparian habitats, marine wetlands, and vacant or ugly urban spaces, will increase as human populations and land prices rise. Ecological, horticultural, design, and engineering professions need to collaborate in doing basic restoration ecology studies and in field-testing the solutions.

5. The need to link the human and natural ecology of urban areas to amenities, social and economic systems, waste management, traffic, urban renewal, employment, and health will accelerate. For example, the public health and animal health fields, until the 1960s, concentrated their research and clinical efforts on identifying and controlling specific pathogens, such as TB and typhus. The realization that some livestock and human diseases, like coccidiosis or trypanosomiasis, do not behave like diseases in wild animals (Dorney 1969) undermines the classical definition of disease by suggesting that diseases have been created by inadvertent human manipulation of the animal-parasite natural environmental interface. Furthermore, the concept of landscape epidemiology (Pavlovsky 1966) demonstrates how certain geographic zones are more liable to transmit diseases to man. The psychological aspects of density (Calhoun 1962) introduced another aspect—crowding—into the public health arena. Understanding disease in an evolutionary, spatial, and density context shifts the emphasis of disease prevention and control from the germ theory to an ecological or environmental focus.

6. The importance will be better appreciated of adapting technology to solving environmental problems, rather than of creating new environmental problems inadvertently with new technology (often under the guise of environmental improvement), such as the detergent fiasco of replacing water-polluting phosphates with NTA (nitrilotriacetic acid), which was found to cause birth defects (Commoner 1971, pp. 156–158). Such technology assessment requires considerable ecological advice.

7. The importance for assessing environmental consequences of devel-

opment activities (ie environmental or biophysical impact assessment, social impact assessment, and risk assessment) will be described in more detail.

8. The need will grow for more applied research on environmental planning and protection techniques, especially costs and benefits with hard numbers.

9. Recognition of the importance of rural and urban land use planning and land use controls will reduce the need for energy-consuming transportation; establish minimum caloric, protein, and energy flows to sustain urban society (the ecobase concept of Caldwell 1970, p. 97; Dorney and Hoffman 1979); and prevent or reduce inadvertent damage to natural resources on and off site.

10. New training centers will be required specifically designed for producing environmental managers or human ecologists (see Chapter 8 on manpower).

11. The need will be perceived to assess the environmental consequences of new and existing legislation.

12. Assisting investors in selecting land for purchase and in selecting investment priorities in harmony with principles of environmental management is important. Prepurchase investigation is vital, because there may be conditions where contaminants from past industrial activities can cost millions of dollars to clean up.

13. The desire of nongovernmental organizations (NGOs) to participate in decisions affecting environmental quality necessitates a two-way information flow of conceptual and technical information. Public participation is closely linked to environmental management (Elder 1975). The environmental manager is a key to coordinating the information packages and maintaining credibility.

All these 13 priority areas take considerable professional manpower. This includes not only environmental managers, drawn from the natural, geographic, engineering, and design areas, able to deal with the human ecosystem in a holistic way, but specialized scientists, designers, and engineers with their nonprofessional counterparts (technologists, plant operators, etc.). Employment projections for environmental scientists/ engineers/technologists were given as between 12 and 100 percent in this past decade (Fanning 1971; Miller and Malin 1972). This figure suggests that if we are to avoid more ecological backlashes from technological fixes proposed by various technically based professionals who have rationalized the problem into simple components, we have to train synthesizers, or ecosystem/systems analysts, who understand that the sum of the parts does not describe the behavior of the whole.

Hence, I believe there now emerges not only the need for identifying a new group of professionals equipped to deal with ecosystem issues in the industrialized countries—people who can tie it together conceptually—

but also a growing demand for such environmental managers. The specialists in fields such as hydrology, botany, zoology, agrology, and civil engineering do not need to be replaced; they will always be needed to provide specific analyses. However, these analyses must be combined with a systems approach taking account of their associated ecological processes. Such environmental professionals having both depth and breadth can be trained by universities and by on-the-job experience, rather than being essentially self-trained, as is the case at present.

Justification for This New Profession in Canada

Canada differs in certain ways from other industrial and nonindustrial countries in the world regarding what I see as its need, and hence its justification, for such a new environmental profession. The specific needs, other than those outlined above, are based on the Canadian cultural reality, its adjudication of powers under the new constitution, its unique geography, and its particular stage of economic development. They are as follows:

1. The need is acute for more fundamental and applied work on arctic ecology. The acceleration in prospective pipeline construction and in extraction of mineral and energy resources requires both basic research on the arctic ecosystem and considerable applied ecology concerning environmental impact of such developments on native peoples. Environmental experience from European and Asian countries can be better utilized than has been the case up to now, but it needs to be put into a Canadian focus.
2. Impacts between urban growth (direct and indirect effects) and top-quality agricultural lands in the Fraser Valley of British Columbia, southern Ontario, and southern Quebec are causing, and will cause, considerable financial diseconomies unless managed in a holistic way. The negative impacts of urban development on agricultural lands and rural lifestyles are balanced at least in part by stabilizing economic factors of urban-based employment for rural residents, and investment in agricultural infrastructure and in increased production per acre. These stabilizing and destabilizing factors need equal research attention, devoid of hysteria (Dorney 1987).
3. Residuals and energy conservation problems are pressing in the industrial urban areas of St. Johns, Montreal, Toronto, Hamilton, Sarnia, Windsor, Sudbury, Calgary, and Vancouver. Toxic waste, unless better managed, threatens health and urban property values. For example, lead levels in urban children may be high (Wallace and Cooper 1986).
4. Technology assessment needs to be done for goods and products

developed in, and imported from, foreign lands and not necessarily adapted to the Canadian culture or its highly variable physical environment.

5. The need to provide for increasing local, national, and international recreational land use without destroying vital ecosystem functions is a continuing issue, especially on northern lakes having limited productivity. The advent of acid rain accelerates further a decrease in lake productivity. Recreation and tourism are major Canadian economies.

6. Major contamination of the Great Lakes by phosphorus, PCBs, mirex, mercury, dioxin, and other industrial contaminants and agricultural pollutants requires continued attention. Remedial action programs on contaminated harbors need to be designed, implemented, and monitored.

7. The impact of air pollutants such as ozone, sulfur, and nitrogen oxides generated from Canadian and U.S. sources on crop production, forests, and lake acidity in Ontario and Quebec requires monitoring and possibly hard international bargaining.

8. The evolution of a landscape mosaic, derived from a colonial past that focused on production, needs to move to a modern, humanistic one of a sovereign people balancing productivity with quality.

Justification for This New Profession in the United States

Many of the special needs in Canada are dissimilar to those of the United States because of differences in cultural attitudes, political structures, geography, climate, and scale of economic resources available. Those professional needs particularly relevant to the United States and likely to cause a demand for more ecological talent there are the following:

1. Assessment is needed of the overseas activities of multinational corporations as they affect the stability of ecosystems in foreign lands (Toffler 1975). Since funding for industrial development often comes from private and/or public funds, such as the Agency for International Development (AID), the Bureau of Institutional Development (BID), the International Monetary Fund (IMF), and the World Bank, professional environmental advice at source is not only ethically vital but politically responsible.

2. Assessment is needed of the impact of existing and new technology developed by government and industry. This assessment is crucial to world ecosystem stability, since so much of the world's new technology is United States generated. Examples of such new technologies are those for weather modification, geothermal energy, fusion power, biotechnology, nuclear waste disposal, and security of computerized natural resource data.

3. From a human carrying-capacity point of view, preparation of plans may be needed for possible redistribution of urban populations in large bankrupt Eastern and Midwestern cities to satisfactory sites located in the Midwest, the mountain states, and the South. This relocation may prevent collapse of estuarine and continental shelf ecosystems now under severe stress because of continued urbanization on the seaboards. Such new settlements could reduce substantially the energy cost of transporting foodstuffs from the grain-producing Midwest and new fossil fuel energy resources from the Rocky Mountains to the East Coast. (Similarly, population relocation could decentralize Canada by its growth from Ontario to Alberta and Saskatchewan.)
4. As for Canada (see 7 above), residual and energy conservation, recreation, Great Lakes contamination, and air pollution effects on crops, forests, and lakes need careful monitoring.
5. Assessment is required concerning water diversions from the Great Lakes to augment water supplies in the Rocky Mountain and plains states. Global warming and increases in evapotranspiration are potential effects.
6. Impacts on coastal development of predicted increases in ocean levels resulting from global warming trends require evaluation.

Justification for This New Profession in Developing Nations

Although there are wide differences between tropical countries, not to mention the subtropical and temperate ones, the developing countries have certain commonalities where establishing a profession of environmental management would be helpful, and may tilt the balance favorably toward political stability and survival of millions of people. Nine specialized justifications for environmental managers include the following:

1. Study is needed of the impact of rural migrants on cities as it affects health of all urbanites, quality of water resources, and loss of natural open space. Vegetation growing around cities in open-space areas, which originally served a protective function like slope stability, now is being overutilized as fuel for cooking and heating.
2. Control in rural areas of fuel cutting and of agricultural or grazing encroachment on unsuitable soils or forest/chaparral vegetation is needed (Naveh and Lieberman 1984). Energy from biomass is an alternative energy fuel source that may be available for cooking and transportation purposes.
3. Assessment and control of industrial pollutants affecting critical natural resources sustaining rural people are needed so that these food and fuel resources are not lost. For example, destroying a fisheries

resource with a toxic chemical waste eliminates a critically needed source of human protein. Since many less economically advanced countries have little or no pollution abatement technology in place, such environmental expertise must be provided as a complete package financed by international banks, international corporations, and internal sources. Furthermore, many foreign firms send to foreign subsidiaries outmoded machinery having no or little pollution abatement equipment built in and install it in countries having little or no pollution legislation, enforcement agencies, and engineering expertise in these matters.

4. Utilization of native plant and animal species for genetic improvement or propagation purposes is needed rather than of non-native biota. In theory, native biota should require fewer pesticides and have less need for veterinary drugs. Use of native African ungulates as a protein resource in tsetse zones rather than susceptible cattle is a case in point.

5. Marginal land management and planning are needed to develop tourism, to develop supplementary biological materials, and to conserve water, soil conservation, wildlife, and national parks (Dr. Thane Riney, personal communication).

6. Urban agriculture and agroforestry offer potential to increase food production where the demand is most intense (Dorney 1983).

7. Natural and earth science mapping at appropriate scales is needed for utilization of land use planning agencies. This includes maps of existing land uses and of land use potential for productivity and urbanization.

8. Establishment of schools of natural resources, planning, environmental management, and environmental engineering needs to be accelerated.

9. Development of new labor-intensive or intermediate-level technology rather than energy- and capital-intensive technologies requires attention. These approaches must be sensitive to local cultures, economic needs, and ecosystem integrity.

Viewed in total, the justification for world environmental management manpower is strong, in fact crucial, for the survival of many countries and societies. If business and technology continue as usual, collapse of many local, regional, and national economies can be predicted as life-supporting ecosystems require more and more shoring up from without in a world having less and less to share. Haiti is a classic case in point; the continent of Africa is particularly vulnerable in this regard. Those countries far-sighted enough to allocate manpower into the ecological sciences should be able to adjust more readily to the "ecospasms" predicted by Toffler (1975). Concurrent with this justification for ecological manpower will come education and professionalization of the manpower.

Definitions of Titles Used in the Environmental Field

The traditional disciplines of engineering, planning, agrology, forestry, architecture, and landscape architecture all have professional societies, nationally and internationally, to perform the role of defining who is, and who is not, a member, and the spectrum of work that falls within their purview. Commonly, state or provincial legislation circumscribes professional identity of these groups for reasons of public protection, restricting who can practice and who can issue licenses.

Many other university-trained men and women have no professional societies, only academically based learned societies or associations with open membership. These include biology, ecology, geology, geography, soils, sociology, political science, psychology, public administration, and wildlife management, among others. These academic areas contain many potential and practicing environmental managers as well. In fact, the Ecological Society of America (ESA), after refusing to discuss joint certification with the Wildlife Society in 1971, now certifies those members who desire such recognition (Ecology Society of America 1983). The Wildlife Society also has its separate certification process (Wildlife Society 1983). Both societies require practical experience to quality for certification; individuals of the core membership, however, are not certified "professionals"—a kind of hybrid situation.

Confusion arises when the descriptive prefix "environmental" is added to the various fields studying the environment. Hoping to clarify the issue, but no doubt not satisfying everyone, I will define various professional titles currently in use. These definitions are at variance with those of Fanning (1971), suggesting that professional titles are still the subject of considerable variation and interpretation; further evolution will occur. Part of the confusion in terminology reflects the diverse origins for this body of environmental professionals and the short span of time (15 to 20 years) that has transpired for this knowledge and practice to develop.

The following definitions of those working in the "environmental field" are presented not to confuse the reader but rather to identify alternative professional terms and their origins. This lexicography will assist in more precisely identifying the origin of the term *environmental manager* and reasons this generic professional title is preferred for the field of activity. A professional normally holds academic qualifications by education and training.

Ecologist: A natural scientist by training who studies ecology from either a synecological or autecological point of view without particular concern for its application. Basically this title distinguishes the theoretician or "pure" scientist from the practitioner or consultant; it includes persons in the employment of both universities and government working for a salary, as distinct from a fee.

Agrologist: Professional agriculturalist involved with animal, soil, and plant production relationships.

Applied Ecologist: An ecologist generally trained in biology either in an academic, consulting, or management role using her or his expertise to solve problems in economics, politics, and logistics (De Santo 1978). This term now is not commonly encountered but rather has been merged with the term *professional consulting ecologist.*

Bioengineer: A person familiar with engineering works incorporating biological surfaces in the design for cost saving or aesthetic purposes (Vanicek 1977; Schiechtl 1980).

Biogeographer: A geographically trained scientist oriented toward spatial description of biotic features, usually botanical (but not necessarily so). Natural resource and park management becomes closely allied to this field, especially in Europe.

Conservationist or Naturalist: A person with unspecified academic or technical training interested in the study, preservation, or management of biotic resources.

Consulting Geographer: A little-used term but, as described by McLellan (1983), equivalent to environmental manager.

Ecographer: An individual is an interdisciplinary "beachhead in a critical no-man's land between two cultures of science"—ecology evolving from the biological sciences and geography from the social sciences (Hafner 1970). This term first proposed by Hafner is similar to the definition I will propose for environmental manager except that I have included in the definition of the environmental manager the additional caveat that he or she have an understanding of how the earth scientists, the design, and engineering professionals, and lawyers carry on their academic and professional activities so that effective communication is facilitated between them and the social scientists and biologists.

Environmental Administrator: As defined by Caldwell (1970, p. 16), an individual who deals with environments holistically by examining *policies, methods,* and *processes* by which humans shape their environment as well as the *control* of human action regarding environmental quality maintenance. The aspects of policy, methods, and processes include both environmental action planning and protection aspects.

Environmental Auditor: A group of professionals approving legal environmental compliance for a corporation, often for its directors and shareholders. It parallels a financial audit. The need for such audits resulted from large court settlements for pollution damage, such as from the Love Canal controversy in Niagara Falls, New York.

Environmental Coordinator: A term used by Royston and Perkowski (1975) which is less forceful than environmental manager. This term is gaining currency for municipal staff positions where integration between departments occurs on issues relating to the environment.

Environmental Designer: A person primarily with design training (architecture, landscape architecture, or planning) who utilizes for design purposes scientific information on physical and natural environmental features and their dynamics.

Environmental Engineer: A person with technical engineering skills who utilizes systems or ecosystems concepts and guidelines for construction or design purposes. Space Engineering (see Odum 1963), sanitary engineering, some aspects of civil engineering, public health engineering, noise control, risk assessment, and pollution abatement activities are covered under this title.

Environmental Lawyer: A trained legal professional who emphasizes environmental quality issues in her or his practice.

Environmental Manager: This is a generic description of a systems-oriented professional with a natural science, social science, or, less commonly, an engineering, law, or design background, tackling problems of the human-altered environment on an interdisciplinary basis from a quantitative and/or futuristic viewpoint. For putting environmental management into an operational mode, two phases are proposed: *ecoplanning** and environmental protection with various separate modes or subsets in Figure 1-1. The modes vary depending on whether the work is oriented towards a development project or towards policy formulation. This definition of an environmental manager is similar to that of ecographer or to the term environmental coordinator.

"Environmental manager" also defines a bridge profession in the sense that those identified as environmental managers commonly hold professional credentials in planning, landscape architecture, engineering, or law. This definition generally follows the one developed by Garlauskas (1975), Petak (1981, p. 214), and Nelson (1982), which suggests that the environmental manager coordinates social, administrative, political, legal, and economic concerns relating to ecosystem management. Although their definitions are more conceptual, the one given here is more operational.

Environmental Mediator: A professional attempting to reconcile conflicting points of view on a particular environmental issue (Dorney and Smith 1985).

Environmental Planner: An environmental planner putting his or her knowledge of the ecosystem into a planning process or predictive frame of reference to effect a better fit between the works of humans and nature. This type of planning is a paper exercise that begins a development process, a government process, or policy formulation process. Environmental planning includes goal setting, information analysis, hearings, and approvals; in this context it is synonymous with environmental management by MacNeill (1971, p. 5), while "environmental planner" according to Glover (1974) and Fabrick and O'Rourke

---

* See page 63 for definition of this new term.

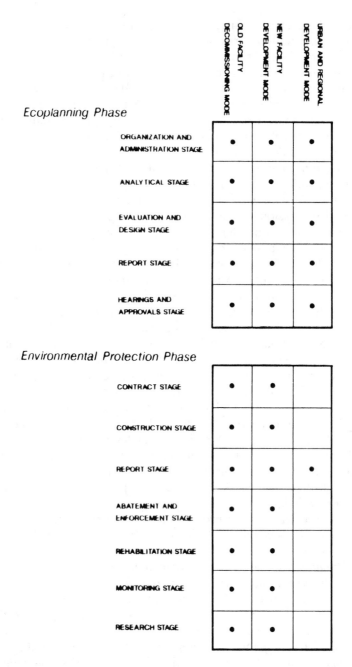

Figure 1-1. The environmental management process broken into phases, modes, and stages.

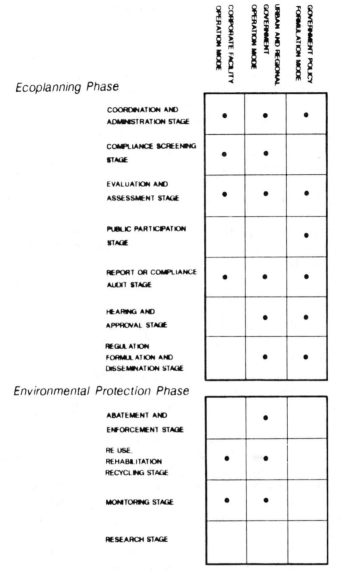

Figure 1-1. *Continued.*

(1982, p. 6) is equivalent to my term environmental manager. As used by Marsh (1983), this term, environmental planner, is synonymous with environmental manager as I have defined it.

Environmental Task Manager: As defined by Fabrick and O'Rourke (1982, p. 100), this is similar to environmental manager, although fewer elements are shown by these authors than are shown in Fig. 1-1.

Environmental Scientist: A science-trained person of many possible subdisciplines oriented toward field studies, not necessarily involving human-dominated ecosystems.

Environmental Systems Analyst: An apt description of the environmental planning aspect but not the environmental protection aspect of environmental management. It appears to offer no advantage over the designation environmental planner, although it has more razzle-dazzle.

Environmentalist: A person of unspecified training concerned about issues of pollution, land policy, land management, or land use planning, often perceived as an environmental advocate. The term may be used in a pejorative way by lawyers, industrialists, and news reporters.

Fishery Ecologist: See Wildlife Ecologist.

Forester: A professional trained in the planning and management of timber crops.

Human Ecologist: A person having either an academic or an applied/ management orientation relating to how human culture and technology affect and are affected by ecosystems in which humans live. It is in essence a human-and-environment focus. The term is broad and appears to defy boundary definition.

Land Use Planner: A person coming from academic planning, geography, the biological sciences, forestry, or environmental studies who conducts, analyzes, and proposes solutions to land use issues, not necessarily with an ecosystem focus.

Physical Geographer: A geographically trained scientist oriented toward spatial description of physical features and natural resources. Generally, the emphasis is toward physiography, geomorphology, or climatology.

Professional Ecologist: A broad generic description proposed by Stark (1973). For this reason I have chosen to subdivide it into three groups: the professional consulting ecologist, the ecologist (mentioned above), and the biological manager—wildlife, fisheries, and (some) foresters. All three of these groups are ecologists; the resource managers are well organized into separate learned or professional societies, such as the Wildlife Society.

Professional Consulting Ecologist: A natural scientist by training who provides for a fee a client advice on ecosystem structure, function, and management. His or her services may include environmental planning and environmental protection activities, but generally not environmental design (see previous definitions).

Senior Ecologist: Term used by the Ecological Society of America to denote ecologists who have the highest credentials for practice (ESA 1983). As well, ESA has the lesser categories of Ecologist and Associate Ecologist.

Systems Scientist: Quantitative analyst, usually a biologist or engineer who deals with information, processes, and decision making in an operational context.

Wildlife or Fishery Ecologist (Biologist): A natural scientist by training interested in faunistic studies and management based on ecological principles, especially on studies of population and sustained
yield.

Professional Integration and Interaction

Since aspects of current planning, economics, landscape architecture,
and architectural and engineering practice interact with those of the
environmental manager, some additional clarification may help at this
point so it is clear how integration without overlap can be achieved. First,
the type of landscape affects the focus of different professional and
academic areas, as shown in Fig. 1-2. For example, design, engineering,
and law concentrate on the right-hand side or urban side of the figure,
while environmental managers concentrate more toward the center. As I
see it, these design, engineering, legal, and social science fields concentrate on urban and regional design, open space studies, urban infrastructure, and land use analyses from six points of view: (1) compatibility
between the physical landscape and the existing and proposed built
environment; (2) socioeconomic values; (3) aesthetics; (4) human uses,
existing or anticipated of the built environment; (5) implementation,
utilizing administration and legal mechanisms; and (6) meeting accepted
technical standards or guidelines for urban construction. Since knowledge
of the natural system, in theory at least, should be in sufficient depth to
understand its subsystem relationships, tolerances, management limitations, and potential integration with the existing and proposed built
environment, it is in working with these complex interrelationships that
environmental managers have a particular advantage.

Environmental managers also have the scientific tools and theories,
based on 50 years or more of ecological field observations and experimentation, to bring these analyses into a problem-solving framework for
improving the fit between humans and nature—that is, a human-oriented
ecology. There is no way the design- or engineering-oriented person can
be expected to be conversant in such technical knowledge; it is difficult
enough for ecologists trained in the sciences to keep abreast of scientific
and technological findings. Indeed, he or she must generally rely heavily
on subspecialists in science, such as archaeologists, anthropologists,
oceanographers, limnologists, entomologists, climatologists, soil scientists, etc. to do so. By way of illustration, Table 1-1 shows the relationship
between level of government and the planning phase of environmental
management, Table 1-2 shows the type of analyses required for a new
town design and planning study, and Fig. 1-3 shows the flow of specialized information by a consultant into an environmental planning and
design process for a new town site (Dorney 1973).

Land Use	Wilderness	National Parks and Forests.	Mixed Recreational/ Agricultural Regions	Agricultural Regions	New Towns and Villages	Urban Areas
Human Density/Acre	None	Seasonal	0.1	0.5	1-20	4-50+
% Natural Landscape	100%	75-98%	25-75%	15-30%	10-30%	0-15%
Mapping Scale	1:250,000	1:50,000	1:25,000	1:25,000	1:10,000	1:5,000
Environmental Management						
Earth Sciences						
Forestry						
Wildlife and Fisheries Management						

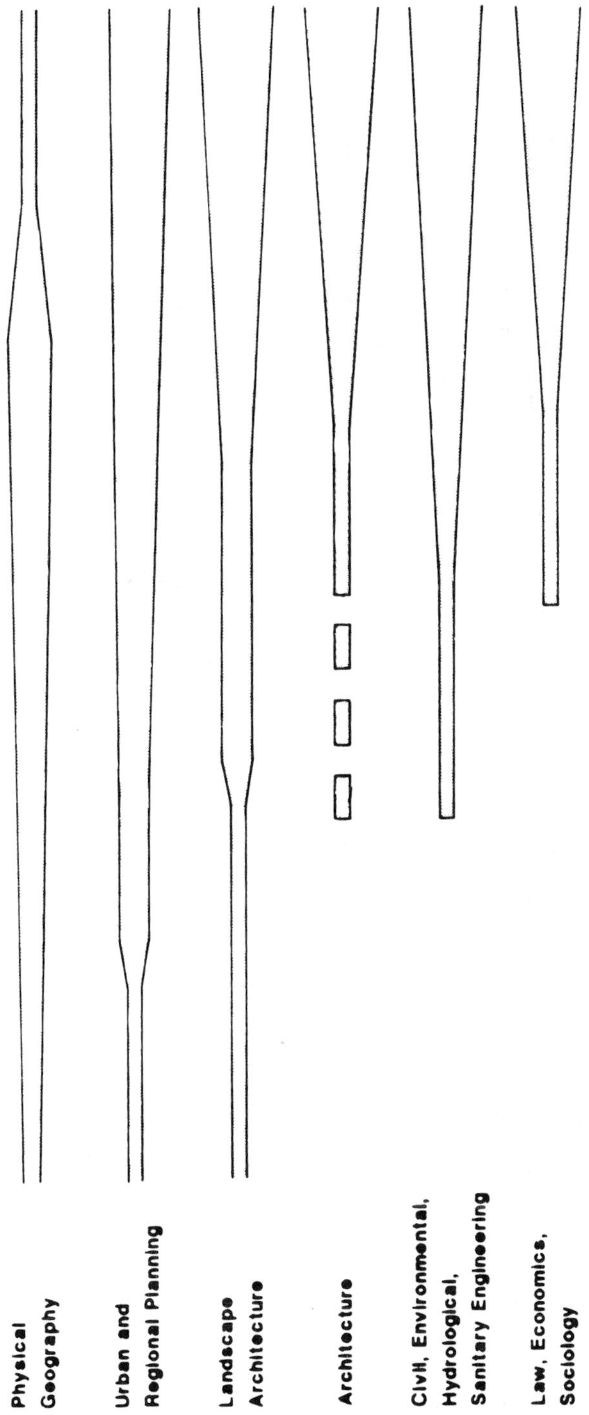

Figure 1-2. A schema of professional interests along a spectrum of land uses, wilderness and urban, and human density.

Physical
Geography

Urban and
Regional Planning

Landscape
Architecture

Architecture

Civil, Environmental,
Hydrological,
Sanitary Engineering

Law, Economics,
Sociology

Table 1-1. Comparative integration of environmental management expertise at five government levels.

Level	Planning approach	Agency organizations	Discipline type
International	Treaties and conventions, conferences, reports, research	United Nations, regional organizations	Environmental management coordination, multidiscipline in focus
National/federal	Legislation, policy, conferences, reports, research, standards, guidelines	Generalized and specialized departments	Same as international, except environmental standards require range of biochemical, engineering, and medical specialties
State/provincial	Same as national/federal, but may include land use designations for major agricultural, mining, and recreation zones	Same as national/federal	Single-purpose resource disciplines plus some environmental management professionals for coordination
County/regional	Resource, biophysical and cultural mapping, land use policy and designations, and zoning bylaws	Planning, recreation, and engineering departments	Planning, recreation, and civil engineering plus environmental management consultants
City	Land use zoning, bylaws	Planning, recreation, and engineering departments	Same as county/regional

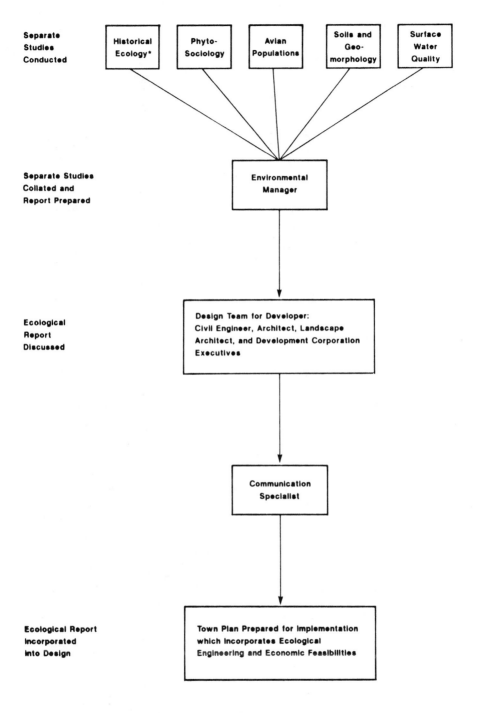

Separate
Studies
Conducted

| Historical Ecology* | Phyto-Sociology | Avian Populations | Soils and Geo-morphology | Surface Water Quality |

Separate Studies
Collated and
Report Prepared

Environmental Manager

Ecological
Report
Discussed

Design Team for Developer:
Civil Engineer, Architect, Landscape
Architect, and Development Corporation
Executives

Communication Specialist

Ecological Report
Incorporated
Into Design

Town Plan Prepared for Implementation
which Incorporates Ecological
Engineering and Economic Feasibilities

* Early vegetation, history of Indian and settlers use of land and water resources, and unique cultural features of the region and the site.

Figure 1-3. Scenario for preparing and independently commissioned environmental report on a new town site (Dorney 1973).

Table 1-2. Breakdown of types of analyses needed and sources of information for ecosystem or ecological analysis.

Discipline	Type of analysis needed	Source of information
Anthropology/archaeology	Location of known sites Location of zones of high probability	Museums, scholars in universities, local clubs, landowners
History or historical ecology	Changes in human settlement patterns in land uses over time Environmental quality indicators Sites of value to residents: Buildings, areas Place names (streets, areas, etc.)	Historical societies, university scholars, archival land use documents (early maps, land surveys, diaries, paintings, photos of landscapes)
Sociology and/or cultural geography	Values and perceptions of residents about present land use and possible future land use	Community leaders, peer groups, questionnaires, hot-line radio and TV shows
Economics	Done by competent specialized firm	N/A
Botany, zoology, forestry, wildlife management	Typical and atypical floristic composition of various physiographic land types Unusual or rare species, biotic communities Forest land capability Wildlife production	Museums, university scholars, government personnel, archives, local naturalist clubs or naturalists, technical publications

Discipline	Data/Analysis	Sources
Geology, soil science	Geologically hazardous landscape formations Land capability for crops Erodibility analysis Mineral/aggregate resources Floodplain zones Groundwater quality, quantity, and depth to water table	University scholars, museums, government personnel and archives, local residents, historical archives Location of existing pits and quarries, well records, boring logs
Limnology and fisheries	Surface water quality Indicator organisms of environmental quality Fishery populations and spawning habitat	University scholars, museum staff and archives, government agencies, local residents
Hydrology	Done by competent separate firm	N/A
Noise	Present noise levels Projected noise levels following construction	Hand-held field measurements, specialized noise consultant or acoustical engineering firm, computer simulation models
Climatology/air	Inversion of potential landscape (mesoclimate) Microclimatic patterns Hills, mountain ranges, and wind direction	Government archives and personnel, university staff, airport records, observations of residents
Snow	Drifting influenced by wind, man-made, structures, and vegetation, natural and planted	Climatology data, model testing in snow simulation chamber usually done by specialized firms
Coastal engineering	Wave energy analysis Littoral drift/sediment transport Erosion	Models of wave size, direction, and probabilities; models of sediment transport/movement; Spartina marsh rehabilitation

The environmental manager, oriented toward the analysis of the human-nature interaction, can serve various professions and politicians by providing coherent analysis of the natural environment in much the same way an engineer calculates traffic flows and produces a hydrology model, and an economic planner calculates the percentage of space in a new town required for a balance between industrial, commercial, and residential uses. Figure 1-4 illustrates this point further for both the planning and protection phases of a facility development, such as for a freeway, pipeline, or waste disposal site. Figure 1-5 illustrates the point for a policy formulation process. Both of these figures show how other professionals' work dovetails with that of the environmental manager and in this sense provides a more realistic overview of team structure and dynamics than Fig. 1-1.

This environmental management focus, then, places the ecological sciences into direct and meaningful interaction with the disciplines of planning, landscape architecture, civil engineering, law, and economics as they relate to urban and regional land use planning, natural resource policy, and allocation or management of lands.

A further way to distinguish the ecological or environmental science-based areas from the more traditional urban design and engineering areas is to view urban development as having four conceptual levels and as requiring four levels of professional involvement (Dorney and Rich 1976).

Level 1: Flat Earth Planning and Design

This level utilizes limited mapping of earth resources. It assumes humans as capable of overcoming any physical limitations or landscape hazards (such as flooding, slippage). Generally either planners, architects, engineers, or politicians acting separately as monument builders make design decisions at this simplistic level of abstraction. Familiar examples of this design approach are early downtown Toronto (laid out in a rigid grid perpendicular to Lake Ontario and running, amazingly, diagonal to river valleys), downtown Washington, D.C., and Brasilia (laid out in the pattern of a bird or airplane).

Level 2: Contour Planning and Design

This level of design and development considers slope and hazard lands before superimposing a human-dominated urban system on the land. Examples are roads running parallel to slopes or at low gradients or parkways paralleling rivers or mountains. Engineers, planners, and architects generally cooperate on projects of this nature.

Level 3: Feature or Constraint Planning and Design

This level of detail requires that critical features be identified, such as historical sites, archaeological sites, unique ecosystems, and hazard lands

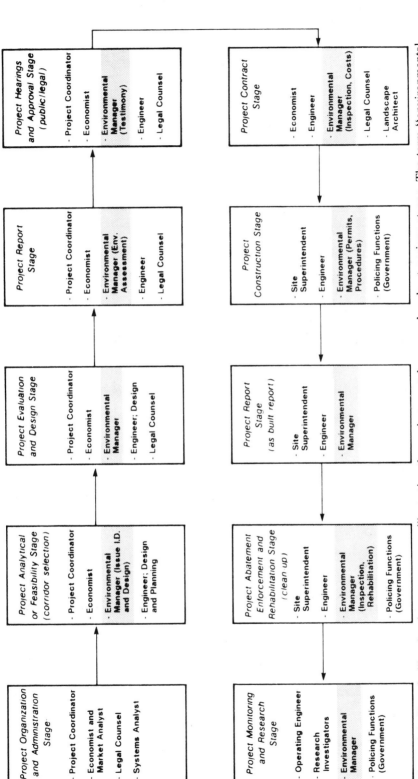

Figure 1-4. New facility development process illustrating the environmental manager's role at various stages. The term "environmental protection" describes the last five stages.

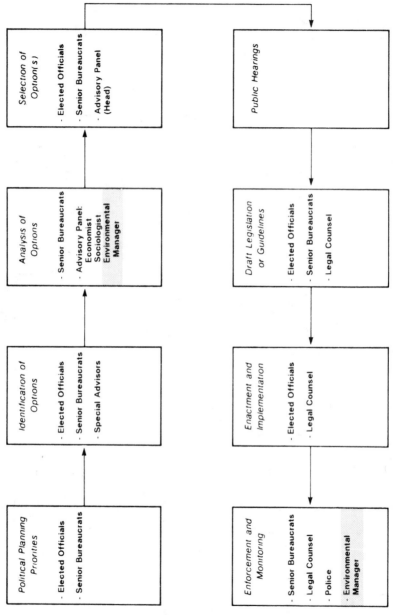

Figure 1-5. The environmental manager's role in two stages of policy formulation.

(floodplains, slopes over 12 percent, etc.) and that they be incorporated into the design. To the professionals identified in levels 1 and 2 are added the landscape architect, archaeologist, historian, physical geographer, geologist, and hydrologist, all working more or less as an interdisciplinary team. The beginning of team effort here is a fundamental distinction from the previous levels 1 and 2, where a single profession dominates the process.

Level 4: Ecoplanning* and Design

At this level of detail, the earth's surface is considered as a dynamic system. In addition to the professional expertise and teamwork needed in level 3, the expertise of the environmental manager is vital. Various subsystem models need to be developed to answer questions of a quantitative and qualitative nature. For example, what water quality can be expected in an urban lake that receives storm water flow from housing of varying densities, and how can quality be optimized?

Generally, most urban design has moved in North America from level 1 to levels 2 and 3 over the past 15 years. Examples of level 4 teamwork that have actually resulted in built environments are few. However, a considerable number have been planned in North America (see i.e., Erin Mills and Townsend new towns, Dorney 1970, 1976a,b; Dorney et al. 1986). Some, like Erin Mills near Toronto and the Woodlands near Houston, are emerging from the cornfields, swamps, and pastureland. Conceptually, this is the design level upon which this book is based.

Distinction Between the Ecologist and the Environmental Manager

This section may seem unnecessary in view of the preceding definitions, yet there is a need to expand on the distinction between ecology as it generally exists among its adherents in universities and in governmental agencies and in the professional practice of environmental management.

The comfort that a guaranteed monthly salary affords if you are a professor of ecology or in a related biological or agricultural discipline enables you to view the business world with its hustle and bustle with some disdain. With the detachment that an academic needs and cherishes strongly built into his or her job, urgent problems rarely exist. Giving answers to difficult environmental questions is usually avoided, since positions based on simplistic analysis might invite criticism from other

---

* The term *ecoplanning*, first used in 1969 by Dorney et al., has been modified to the comprehensive version in Fig. 1-1. It is synonymous with environmental planning as also defined by Land and Armour (1980, p. 20).

ecologists and jeopardize the objectivity or academic freedom of a professor. Whether or not the Ecological Society of America certifies the ecologist does little to change the basic reality—ecologists do not receive training in professional practice and have minimal training in law, ethics, and social and geographic sciences. They are not equipped by virtue of their academic training to be environmental managers.

Not infrequently, however, academic ecologists hold seminars or meetings dealing with issues of environmental quality and management. Such meetings and published proceedings often attempt to establish why management professionals should include ecological information in decision making (Roberts and Roberts 1984; Bliss, President's Address at Ecological Society of America's Banquet 1984) without dealing with the sociopolitical and ethical implications of what becoming a professional practitioner requires. Such occasional bouts of chest thumping belie the need for structural changes in academic ecology, if it is to be useful to decision makers.

The environmental manager, as I know the situation, is projected generally into the hub of the economic and political spinning; he or she may often be brought in after an environmental debate has begun, and advice is often only intuitive or based on reconnaissance-level investigation. Time, and to some extent money and manpower, may not always be available for complete documentation or analysis, which increases the probability of error. The environmental manager must be able to gauge the probable extent of error resulting from insufficient understanding of the ecosystem and so advise a client. In addition, if the issue is likely to go to public hearing or to the courts, it is vital that as a professional, issues of confidentiality are clearly identified vis-à-vis a public or legal inquiry, along with the ethical issues surrounding public disclosure, accurate testimony, and credibility.

An environmental manager, employed by a government agency or a nonprofit institute, is likely to be engaged in developing environmental policy and environmental legislation, coordinating of environmental affairs with other agencies, assessing environmental impact statements, carrying on technology assessment, conducting research, or undertaking public education or policy studies. If he or she does not take a fee for the work but is on salary, a true client-practitioner relationship may not exist—a sharp and necessary distinction. Furthermore, the public and interdepartmental pressure generated by particularly controversial issues may be intense.

The attempt by the Ecological Society of America to certify ecologists for practice is a partial attempt to paper over the chasm between the academic and the practitioner. Its flaws are three: first, by refusing dialogue with the Wildlife Society for joint certifications it allowed the National Association of Environmental Professionals (NAEP) scope for its evolution; second, by being an international learned society, it cannot certify other nationals for political-legal reasons; and third, with academic dominance of the membership, it precludes the likelihood of professional debate and professional evolution. The momentum for professional development for these structural and jurisdictional reasons is with NAEP.

Distinctions Between the Environmental Manager and the Agrologist, Forester, Wildlife Manager, and Physical Geographer

Understanding the human-dominated ecosystem sufficiently well to provide definite answers for businessmen, planners, administrators, and legislators requires that information provided by or available from the natural resource managers (soils, forestry, wildlife, fisheries, etc.) be integrated correctly. An example might be helpful in distinguishing how this would work. To understand the significance to environmental quality of nitrogen and ammonia sources (residuals) generated in an urban and dairy-farming region requires information on dairy cattle numbers, manure and fertilizer use, soils, slope, crop cover, amount of automobile exhaust, and groundwater, stream, and lake water quality monitoring.

From this compilation, coupled with chemistry of rainwater and well water, an ecological assessment of the relative importance of agricultural and urban waste residuals might emerge. The potential impact of these residuals on forestry, wildlife, crop production, fisheries resources, and human health (important from a contaminated well water point of view) could then be determined. The advantage in planned intervention or systems management, if any, is determined and then followed by implementation as required. The environmental manager is responsible for coordination of environmental science inputs; the resource managers contribute their specialities. Parenthetically, this distinction between the two areas of contextual and specific knowledge is not shared by Alston and Freeman (1975), who see the natural resources manager as handling both types of information.

Hence, from this example, the difference between the term environmental manager as I use it and the more specialized resource manager with or without certification is one of the level of detail, the level of systems integration, and the human ecology focus. These are not at all minor differences. Credibility for the environmental manager requires being more or less abreast of current concepts and field practices in soils, forestry, wildlife, and fisheries management along with some knowledge of institutional and social systems. This means attending various meetings, reading core journals, and working shoulder to shoulder with resource specialists. Yet, the competence to integrate these separate science inputs in a decision-making context is a crucial component beyond having a grasp of technical information and concepts.

Another difference between the environmental manager and the other resource specialists can arise from the level of detail needed to answer the problem at hand. For example, to a wildlife biologist, a mammalian inventory may include everything from deer to mice; the environmental manager may be content with an inventory of ungulates and carnivores, only assuming that carnivores would not be present, such as the least weasel, unless mice were too. I have seen a new town environmental assessment project (Pickering, near Toronto) where mice were inventoried and tabulated by the wildlife biologist, who insisted that this was the only way he could do a professional job. Later the newly hired ecological consultants found no use for such mouse data. In terms of community design, such information was not only wasteful but useless. Such wasteful field science can only be avoided by ensuring that the resource manager's inputs are carefully tailored or controlled at the outset of the project. Unless this is done, it can cause other professionals to ridicule "ecological" advice, such as in this mouse/new town design example.

Similarities between all the resource management fields (including physical geography) and ecology relate to similar basic undergraduate training and course work as well as to certain common environmental planning principles, which are elaborated more fully in Chapter 2. These similarities generally provide a common bond upon which strong personal and professional relationships and mutual trust can emerge. On the other hand, the absence of common undergraduate background, the lack of a common technical language, and lack of familiarity with planning concepts and principles among the environmental manager, the engineer, and the design professional, discussed further in Chapters 2 and 3, are obstacles to effective communication, a situation not present between ecologists, resource managers, and physical geographers.

Trying to be specific about how the physical geographer fits into this professional resource management arena is not easy, since the discipline of geography is understandably broad. Some physical geographers could quite easily be described as environmental managers, playing the role of synthesizing technical information (McLellan 1983). Other physical geographers more nearly approach the role of specialists in soils, geology, or climatology. Hence, any generalization about geographers is best avoided, each being individually labeled as the need arises.

The analytical data on which the separate resource managers and environmental manager depend are similar. For a comprehensive discussion of landscape and natural resource analysis techniques at aerial photo scale, Marsh (1978) should be read, or Lang and Armour (1980). The essential point, however, is that the data assemblages done by each discipline are not necessarily coherent or organized, unless both a conceptual and technical framework is first developed. This organizational framework development falls to the environmental manager in the ecoplanning phase (Fig. 1-1).

Distinctions Between the Environmental Manager and the Environmental Engineer, Landscape Architect, Environmental Planner, and Environmental Lawyer

The schools of planning and landscape architecture in Canada and the United States, and to a lesser extent in Western Europe, have broadened their focus from policy, administration, and design (aesthetics, project management, specification writing) to broader environmental issues. This began with the hiring of ecologists (natural resource managers) in the late 1960s. McHarg's *Design with Nature* (1969) accelerated the academic change and opened up a professional dialogue. Law schools responded by an interest in environmental law and ethics; engineering schools responded by developing environmental engineering, systems design, and management science departments. By the end of the 1970s, considerable interest in ecosystems, particularly for regional landscape issues, was embedded in these older discipline areas.

The environmental manager, as the term is used and developed in this book, is not a landscape architect, a lawyer, a planner, or a civil engineer, but a landscape architect, lawyer, planner, or engineer may choose to become an environmental manager. The matter is one of focusing on a systems approach, broadening academic interests, and developing information-coordinating skills. In this regard environmental management is a bridge profession tying ecosystem knowledge to these more classic professional roles. In Canada many members of the Ontario Society for Environmental Management (OSEM) are members of the Canadian Institute of Planners (CIP), the Ontario Society of Landscape Architects (OSLA), the Ontario Bar Association, and the Association of Professional Engineers of Canada; they are dual citizens, so to speak. Ideally, one might wish that environmental management could emerge full blown from academia with a separate program and clear mandate, but this has not happened. Nonetheless, the bridge professional approach and identity have worked. This book will assist in its further evolution rather than belabor the point that while ecologists may have invented the ecosystem, human ecology was beyond their academic and conceptual grasp.

Organization of This Book

In the following chapters I describe the philosophical and technical principles for a professional practice, the conceptual basis for environmental management, the ways to organize a practice, the client-practitioner relationship, the preparation of technical reports, the costs versus savings for ecological studies, the training and professionalization in environmental management, and finally future prospects for the environmental manager.

Organized in this way, I hope the reader, whether student, ecologist, resource manager, designer, engineer or lawyer, can find areas of interest to him or her. Because each profession develops a body of knowledge it holds sacred, my biases are easily identified as coming from the natural sciences.

As more experience develops in environmental management practice, it is likely some formal national or federal professional structure will arise in Canada and the United States, giving us an annual forum for debate and discussion of the kind of issues brought forth in this book. To date OSEM is one of the first such professional bodies, but it only covers the Province of Ontario. The NAEP in the United States is another. Until more state, provincial, and national groups are organized, informal contact provided by books such as this, by symposia sponsored by other professional and various learned societies, and by the Institute of Ecology will have to suffice.

Bibliography

Alston RM, Freeman DM (1975) The natural resources decision-maker as political and economic man: Toward a synthesis. J Environ Mgmt 3:167–183.

Bacon EN (1967) Design of Cities. New York: Viking Press.

Bliss LC (1984) Ecologists' need to increase their involvement in society. ESA Bull 65(4): 439–444.

Boyden S, Millar S, Newcombe K, O'Neill B (1981) The Ecology of a City and Its People: the Case of Hong Kong. Canberra: Australian National University Press.

Caldwell LK (1970) Environment: a Challenge for Modern Society. Garden City, NY: Natural History Press.

Caldwell LK (1982) Science and the National Environmental Policy Act. University: University of Alabama Press.

Calhoun JB (1962) Population density and social pathology. Sci Am 206(2):139–148.

Commoner B (1972) The Closing Circle: Nature, Man and Technology. New York: Alfred A. Knopf.

Darling FF, Milton JP (eds) (1966) Future Environments of North America. Garden City, NY: Natural History Press.

De Santo RS (1978) Concepts of Applied Ecology. New York: Springer-Verlag.

Dorney RS (1969) Epizootiology of trypanosomes in red squirrels and eastern chipmunks. Ecology 59(5):817–824.

Dorney RS (1970) The ecologist in action. Lands Arch (April) 196–199.

Dorney RS (1973) Role of ecologists as consultants in urban planning and design. Hum Ecol 1(3):183–200.

Dorney RS (1975) Incorporating ecological concepts into planning, with emphasis on the urban municipality. In: Nature and Urban Man, Special Publ No. 4, Canadian Nature Federation Conference, Dec., pp. 99–110.

Dorney RS (1976a) Incorporating the natural and historic environment into Canadian new town planning. Contact 8(3):199–210.

Dorney RS (1976b) Biophysical and cultural-historic land classification and mapping for Canadian urban and urbanizing land. In: Ecological (Biophysical) Land Classification in Urban Areas, Ecol Land Class Series No. 3. Ottawa: Environment Canada, pp 57–71.

Dorney RS (1983) The urban ecosystem: Its spatial structure, its scale relationships, and its subsystem attributes. Paper presented to World Systems Conference, Caracas, Venezuela, July 10–16.

Dorney RS (1987) Urban agriculture in southern Ontario: Myths, realities, options. Environments (forthcoming).

Dorney RS, Evered B, Kitchen CM (1986) Effects of tree conservation in the urbanizing fringe of southern Ontario cities: 1970–1984. Urban Ecol 9:289–308.

Dorney RS, Hoffman DW (1979) Development of landscape planning concepts and management strategies for an urbanizing agricultural region. Lands Plan 6:151–177.

Dorney RS, Rich SG (1976) Urban design in the context of achieving environmental quality through ecosystems analysis. Contact 8(2):28–48.

Dorney RS, Smith LE (1985) Environmental mediation. Working paper No. 19, University of Waterloo (Ontario) School of Urban and Regional Planning.

Ecological Society of America (1983) Registry of certified ecologists. Bull Ecol Soc 64(1):32–36.

Elder PS (ed) (1975) Environmental management and public participation. Toronto: Can Environ Law Res Foundation and Can Environ Law Assoc.

Fabrick MN, O'Rourke JJ (1982) Environmental Planning for Design and Construction. New York: Wiley-Interscience.

Fanning O (1971) Opportunities in Environmental Careers, Vocational Guidance Manuals. New York: University Publishing.

Garlaukas AB (1975) Conceptual framework of environmental management. J Environ Mgmt 3:185–203.

Glover FA (ed) (1974) Environmental impact assessment: A precedure for coordinating and organizing environmental planning. Tech Publ No. 10. Boulder, CO: Thorne Ecological Institute.

Hafner EM (1970) Toward a new discipline for the seventies: Ecography. In: Mitchell JG, Stalling CL (eds) Ecotactics: The Sierra Club Handbook for Environmental Activists. New York: Pocket Books, pp. 211–219.

Holling CS (ed) (1978) Adaptive Environmental Assessment and Management. New York: Wiley.

Hough M (1985) City Form and Natural Process. New York: Van Nostrand Reinhold.

Lang R, Armour A (1980) Environmental Planning Resource Book. Ottawa: Lands Directorate Environment Canada.

MacNeill JW (1971) Environmental Management. Ottawa: Privy Council Office, Government of Canada.

Marsh WM (1978) Environmental Analysis for Land Use and Site Planning. New York: McGraw-Hill.

Marsh WM (1983) Landscape Planning: Environmental Application. Reading, MA: Addison-Wesley.

McHarg IL (1969) Design with Nature. Garden City, NY: Natural History Press.

McLellan AG (1983) The geographer as practitioner: The challenges, opportunities, and difficulties faced by the academic consultant. Can Geog 27(1):62–67.

Miller S, Malin M (1972) Jobs in the environmental field. Environ Sci Technol 6(8):694–699.

Naveh Z, Lieberman AS (1984) Landscape Ecology. New York: Springer-Verlag.

Nelson JG (1982) Public participation in comprehensive resource and environmental management. Sci Public Policy (Oct.), 240–250.

Odum HT (1963) Limits of remote ecosystems containing man. In: AE Lugo, SC Snedker (eds) Readings on Ecological Systems: Their Function and Relation to Man. New York: Manuscripts Educational Publishing, pp. 316–330.

Odum HT (1983) Systems Ecology. New York: Wiley.

Papageorgiou JC (1980) Management Science and Environmental Problems. Springfield, IL: Charles C. Thomas.

Pavlovsky EN (1966) Natural Nidality of Transmissible Diseases with Special Reference to the Landscape Epidemiology of Zooanthroponoses. Urbana: University of Illinois Press.

Petak WJ (1980) Environmental management: A system approach. Environ Mgmt 4(4):287–295.

Petak WJ (1981) Environmental management: A system approach. Environ Mgmt 5(3):213–224.

Reekie F (1975) Background to Environmental Planning. London: Edward Arnold.

Roberts RD, Roberts TM (1984) Planning and Ecology. London: Chapman and Hall.

Royston MG, Perkowski JC (1975) Determination of the priorities of 'actors' in the framework of environmental management. Environ Conserv 2(2):137–144.

Sarnoff D (1967) The new profession—manager of environmental forces. Environ Sci Technol 1(11):887.

Schiechtl H (1980) Bioengineering for Land Reclamation and Conservation. Edmonton: University of Alberta Press.

Spirn AW (1984) The Granite Garden: Urban Nature and Human Design. New York: Basic Books.

Stark N (1973) The profession of ecology. Bull Ecol Soc Am (Summer), 4–5.

Toffler A (1975) The Eco-Spasm Report. New York: Bantam Books.

Vanicek V (1977) Eco-engineering—an ecological approach of land reclamation and improvement to landscape environment. Lands Plan 4:73–84.

Wallace B, Cooper K (1986) The Citizens Guide to Lead. Toronto: NC Press.

Wildlife Society (1983) Certification. Wildlifer 197:12.

World Commission on Environment and Development (Brundtland Report) (1987) Our Common Future. Oxford, U.K.: Oxford University Press.

2

Philosophical, Ethical, and Technical Principles for Environmental Management

Articulating a philosophical and conceptual underpinning for environmental management is a challenging task. Certainly, it needs to be done so that the core of the new professional territory is clearly defined. This task has been undertaken in previous papers (Dorney 1977a,b) and is derived as well from dialogue generated by workshops that drafted the Constitution for the Ontario Society for Environmental Management (OSEM) (1976). Experiences from these two efforts are combined here.

Philosophical Principles

Although there may be some disagreement among those who either teach or practice environmental planning, protection, and management, there seems to be at least some appreciation that the land ethic of Aldo Leopold and the reverence for life of Albert Schweitzer are powerful and pervasive statements. Coupling with these the more recent concerns for endangered species, animal protection, natural area preservation, and cultural preservation, a new "ethic of diversity" emerged in the 1960s and continues to the present. The plethora of animal TV shows, the save-the-whale-movements, the anti–seal hunt advocates, and the antihunting and antitrapping groups express this concern for diversity and to some extent with the reverence-for-life concept. The preservation of minority language rights and the attempt to preserve ethnicity are cultural and social expressions of a parallel nature.

This ethical triad for ecology (Dorney 1977a)—reverence for land, life, and diversity—poses some problems for professionals who do not necessarily subscribe to its persuasions. It may pose problems as well for Christian theologians who see some threads of Eastern religions in its tapestry, to politicians relating to the Green movement, and to adjudicators of majority and minority rights.

Obviously many urban or industrial developers who alter landscapes cannot necessarily accept the fact that their proposal may not be acceptable because a rare or endangered species is occupying the site. Yet reconciliation of this reality is being done: witness the snail darter controversy over dam construction in the United States and the West Virginia sulfur butterfly versus Ontario Hydro in Canada. This respect for life and diversity is happening not only in North America but in many countries: sensitive area and endangered species legislation is becoming more common; native or aboriginal rights are being considered prior to project implementation. The word and the concept of "ethnocide" are providing a conceptual middle ground between genocide and social impact.

Toxic waste disposal emerges as a major technical and ethical issue. Commonly, people question the morality of burying waste, especially nuclear waste whose half-life is in the thousands of years. In the workplace, handling of toxics generally must follow regulations established by government. However, an employer has a moral obligation not to knowingly injure or harm employees. Regulations, reinforced by social norms of fair play, are also broadly related to the reverence for life.

Understanding the reverence for land, life, and diversity does not necessarily make it easier for the lawyer, planner, plant foreman, or engineer to act in a responsible way toward employees, clients, and agencies. Identifying and understanding this perspective can assist professionals in assessing differences in opinion and in responding to the reality of a situation as it emerges in public hearings, courts, council meetings, and comments from agencies circulated in an approvals process.

In a slightly different vein, the systems perspective that views natural and cultural landscapes as an integrated and interacting dynamic series of hydrological, climatological, geological, pedological, biological, and cultural-technological subsystems, or minerotrophy, phytotrophy, zootrophy, investment, and control levels (Dansereau 1973), also has altered our response to land use planning. Some architects, planners, designers, and engineers used to, and some still do, conceptualize land areas as a void waiting for "redemptions" that is, for development. However, this flat-earth type of planning has evolved to viewing space in dynamic terms—that is, as an ecosystem. A conceptual evolution and revolution going from flat-earth planning to ecosystem planning, as discussed in Chapter 1 (Dorney and Rich 1976), is occurring. This systems perspective has modified professional practice and academic disciplines in the short space of 15 to 20 years, a remarkably rapid evolution.

Closely related to systems thinking is the development of computer modeling arising from the mathematical sciences (see Boughey 1976). This capability was quickly adapted by ecologists to ecosystem modeling, either for natural areas devoid of humans, such as the IBP Biome models, or for larger ecosystem models including humans, such as the spruce budworm spraying programs in New Brunswick (Holling 1978).

The Club of Rome model is the global systems model that captured the public and professional imagination (Meadows et al. 1972). Although it only demonstrates what could happen, as opposed to what will happen, if population growth, resource consumption, and pollution follow certain trends, it is an example of an unexpected development in one field profoundly affecting public and professional attitudes about technology and the growth ethic. Climatological modeling (U.S. CIA 1976) has also altered our perception of future effects on crop production should global climate become more erratic.

Concurrently, a whole series of more specific air pollution and hydrological models have been spawned to intrigue the planner, resource manager, and consulting engineer. Their application holds much promise but requires a level of professional, public, and political acceptance which cannot be achieved overnight. Federal, provincial, and state strategic planning become important for our collective survival, particularly in the attempt to soften or adjust to regional crop failures.

Since landscape mosaics, or patterns, evolve over time, affecting all parameters of the landscape (diversity, productivity, externalities), the mechanisms causing this change need identification before control and adaptation are attempted. I have suggested that landscape evolution is driven by four principal factors: institutional, social, technological, and economic (Dorney 1985). However, I have added ecology as a fifth factor (to be discussed in Chapter 3). This evolutionary process (also discussed in Chapter 3; Figs. 3-1, 3-2) works both at regional scales, where land use policy, interest rates, and marketing boards affect the land use change, all the way down to the local scales, where zoning changes and landowners' attitudes effect small but cumulative change. The result may be that the landscape system adapts slowly or rapidly. Natural landscapes, agrolandscapes, and urban landscapes all undergo these processes.

Technology as one of the prime factors in landscape evolution can be assessed for its potential ecological effects prior to being manufactured, field tested, or disseminated, but this is rarely done. All societies tend to be captivated by technology and by allied marketing forces promoted by narrowly based academic and industrial research of a reductionist nature funded by industry. Breaking this potentially destructive cycle, without losing the beneficial changes, is not easy; it is not helpful to throw out the baby in the bath water. However, what is required is an environmental technology assessment group able to deal with the processes affecting landscape evolution, not unlike many of the initiatives undertaken by the U.S. Council for Environmental Quality and the U.S. Environmental Protection Agency at the upper level of government with its local counterparts. Then landscape change can be guided in a more thoughtful way rather than in the present wasteful ad hoc process, which seems to be two steps forward and one back.

Whether or not all planners accept this ethical (or ecological) triad as a philosophical basis for planning land, water, and living systems of the Spaceship Earth; whether or not they accept the systems concepts and related computer modeling in their practices to define dimensions of reality; and whether or not the concept of five factors explaining landscape evolution is accepted, the point is that some agencies, public-hearing bodies, and private groups are doing so, in Canada, in the United States, and abroad. It is amazing how rapidly these concepts have been adopted, at least in part, by various professionals. Whether the concepts will be further institutionalized as a political movement, as has happened in Western Europe, or as a new religion of ecology remains to be seen. From a politician's point of view, such a conceptual shift in beliefs and professional attitudes is significant. Such a shift allows the decision maker some room for delaying projects (as was proposed by the Berger Commission [Canada] for the northern MacKenzie Valley pipeline), for phasing others, or for rejecting them altogether using as a justification the ecological or environmental argument. Such a shift also leads to innovative design and management and to an ecologically sound, socioeconomic supporting landscape mosaic—an evolutionary process. Perhaps peoples and governments will see the wisdom in this latter approach.

In all these conflicting historical threads, it is important to emphasize that if we seek answers to problems by examining only technological solutions, we have narrowed the problem substantially. Some technological development issues have social, ethical, and political solutions (Shrader-Frechette 1982). For example, traffic congestions can be corrected by widening a road or by having workers do "slip hours," arriving at work any time between 8 and 9 AM and leaving anytime between 4 and 5 PM, a cheaper solution.

In a slightly different vein, enlightened self-interest offers another philosophical justification for environmental management. When the public or government intervenes in a proposed development, considerable negative effects can be generated. These include delays, skyrocketing construction costs, increasing regulation, plant shutdowns, lost markets, fines, loss of reputation and goodwill, and even civil disobedience (Findley 1981). The response of business is either to fight or alter its planning and operations to include environmental management concerns; the latter is less expensive, in Findley's view, than muddling through. It has also been given the name adequate environmental management (AEM), which means the minimum investment in environmental studies that controls or reduces to a minimum the overall or total cost of the project distributed throughout the facility design, approvals, and construction stages (Loucks et al. 1982).

Conceptually, there seems no quarrel with the fact that long-term economic benefits accrue from environmental management. When a landscape loses its productivity, standards of living are under negative pressure; when a landscape loses its aesthetic appeal, property values diminish. The tension between economies and environmental management usually centers around short-term issues. Chapter 7 addresses both these short-term and long-term issues and demonstrates that positive socioeconomic reconciliation can be achieved if certain procedural adjustments are made.

The traditional regulatory approach for the purposes of effecting a satisfactory outcome in environmental quality has tended to become the usual mode proposed by legislators, bureaucrats, and the public. More recently this single-minded approach is being challenged by other planning processes that mix normative perspectives (what we want to achieve—the goals and objectives) with education and mediation. Thus the philosophical basis for environmental management is broadening for reasons of cost effectiveness and political sensitivity.

In my judgment population policy is an area with which environmental management should not mix. Although it can determine environmental capacity from a land production point of view, the tradeoffs among standard of living, growth, technology, quality of life, and religious beliefs are best left to the policy makers. Some countries have opened up these arguments, for example in Australia (Gilpin 1980): the Australian Conservation Foundation has recommended to the National Population Inquiry a long-range population resource exploitation target (Ibid. p. 110). To me this goes well beyond the scope of environmental management. It belongs in the arena of social values formation and political implementation.

Moral and Ethical Perspectives

Professional moral and ethical perspectives become an issue once a body of knowledge develops some boundaries and individuals "profess" to teach and act within those boundaries. Related to this is the protection of the public from unqualified individuals. This concern tends to evolve to the point that professionals eventually form societies, are legally licensed, develop specific educational curricula, become certified, and exclude lay individuals from offering opinions to the public under a professional label. Environmental management now has evolved so that it does have, in my opinion, a body of knowledge to profess. The technical principles in the subsequent sections of this chapter in part provide that bounding of knowledge.

For the subsequent stages outlined above, that is, for professional evolution to occur, more time and more experience are required (see Chapter 1). Licensing at this time in any jurisdiction is inappropriate. However, the environmental management practitioner now selling, or contemplating selling, professional services to the public and private sector should at least be aware of the following general ethical concerns (drawn in part from the OSEM Code) and seen through the eyes of other professions, such as planning, landscape architecture, and engineering:

1. Recognize responsibility for maintaining environmental quality
2. Recognize that the practitioner has obligations to the larger public as well as to the client
3. Price work in a reasonable and fair way
4. Maintain courtesy to fellow professionals by refraining from criticism unless criticism is done in an open forum allowing for rebuttal
5. Undertake work that replaces other professionals only when the other professionals have been paid in full for their services and all contractual obligations to them are discharged
6. Review of professional misconduct should be done by a duly delegated ethics committee whose authority has been delegated by a professional organization
7. File accurate reports that do not hide or disguise information that if fully disclosed could lead to damages to innocent parties
8. Advertise technical expertise in an accurate manner
9. Exchange information in newsletters and journals
10. Provide counseling services by peers to members confronted with ethical matters
11. Present major public concerns and issues to the profession in a systematic way through meetings and newsletters
12. Interpret laws and regulations accurately to the client
13. Make expertise available when possible on a pro bono (free of charge) basis where a needy client has no funds
14. Publish expertise whenever possible to widen the knowledge base of other practitioners
15. Devote time to educating younger professionals
16. Participate in in-career educational programs to upgrade skills.

From another perspective, nine of these 16 ethical concerns can be viewed in an activity matrix as an interaction between objectives, professional associations, and professional activities (Table 2-1). In addition, some of the above 16 issues (Reynolds 1981) go beyond strictly ethical concerns; that is, they may be considered partially operational in nature. Even though the boundary line admittedly may be ill defined between these two perspectives, the 16 issues cover situations commonly encountered in practice. Furthermore, if violated by a practitioner, they generate criticism by other professionals even if there is no formal protest to an ethics committee (assuming such a body has been formed to hear complaints).

It is important, I believe, that a code of ethics not stifle intraprofessional criticism. Professional practitioners should disagree; the issue of how this is done—in a forum where rebuttal is possible—should not in any way suppress honest disagreements from being presented to the public or being presented before legal proceedings. Done this way, disagreement leads to improvement in practice and in the diminution of malpractice charges rather than the reverse.

Technical Principles for Environmental Management

Underlying any new professional endeavor should be a set of technical principles. Once these have been tested in the social, political, and legal forums, they rapidly become norms to guide and teach future environmental managers, other professions, and the public. These principles can guide requests for proposals, be the basis for testing the adequacy of planning reports, contest and evaluate environmental assessment reports, and suggest new areas for research. Without such guiding principles, the response of the public, the professional, and the politician can only be ad hoc. Surprisingly, only Walker and Norton (1982) have attempted to produce such a list for the science aspects of environment. These ideas, derived from my experience, also build on those of Odum (1969), McHarg (1969), Holling and Goldberg (1971), Mattyasovszky (1975), and Kaiser et al. (1974). An earlier version of these principles was published by Dorney and Hoffman (1979).

Without guiding principles, environmental emotionalism has an open field, and any opposition can make equally irrational comments. It is, I believe, useful to begin codification of principles as they have evolved to this point in the hope that they are useful and encourage further evolution.

Not all the principles are applicable for every project. The scale of the project, type of decision being made, time and cost of the planning problem, and the needs of the client influence the focus and level of detail. However, a set of guiding principles serves as a useful checklist when terms of reference, budget, and time are allocated for study. Guiding principles also assist students in the professional schools of engineering, law, planning, and landscape architecture to understand how environmental planning ties in with their fields. For example, if a matrix is used with the 51 principles on one axis and different technical reports on the other, the key differences and similarities between the reports become apparent.

These principles organized around three thematic areas—general planning, natural science, and social science—are as follows:

Table 2-1. Interaction between professional objectives, activities, and constraints on a professional association (Reynolds 1981).

	Exchanged ideas	Codes of ethics	Counsel members	Examine complaints	Sanction unethical members	Aid, assist members with moral problems	Advocates for foreign colleagues in trouble	Present members' contributions to greater society	Determine relay societal concerns to members
Moral/ethical objectives									
Personal assistance	X	X	X						
Moral homogeneity	X	X	X	X	X				
Collegial advocates			X			X	X		
Standardizing competence		X		X	X				
Member/society interface								X	X
Constraints upon associations									
Societal norms and values	X	X				X	X		
Legal standards for responsibility			X			X			

	1	2	3	4	5	6	7	8
Specific government controls				X	X	X		
Employer influences								
Confidence and support of members	X		X		X	X	X	X
Association relevance to member career interests and advances				X		X	X	X
Legal status of association on access to the profession				X	X			
Tax status of the association		X		X	X			
Legal constraints on association control of membership								X
Impact and influence of association with foreign governments			X					

Principles Applicable to General Planning

1. Identify the *planning process* to be followed (ad hoc, incremental, reactive, rational, mixed scanning, transactive, normative, advocacy, radical).
2. Identify site-specific *environmental goals and objectives* and relate these to broader policies of various government levels to guide the planning process, to promote evaluation, and to facilitate public and interagency communication.
3. Evaluate *new technology* from ecological, natural resource, social, economic, and institutional perspectives.
4. Examine *justification,* both conceptual and technical, for a proposed project.
5. Assess new projects from an *environmental impact* perspective and define the undertaking in precise terms:

   Determine the environmental evaluation process
   Determine direct and indirect, positive and negative, short-term and long-term effects
   Plan timing
   Examine alternative actions, including a no-build alternative
   Identify existing trends without intervention
   Examine mitigation measures and resource losses that cannot be mitigated
   Prepare an evaluation of composite site suitability
   Identify those groups and activities benefiting versus those that do not
   Determine scope and scale of each project phase (acquisition, population relocation, demolition, construction, occupancy, maintenance, mitigation, and decommissioning)
   Develop an environmental protection plan.*
6. Undertake *environmental protection planning* to include mitigating measures, their degree of risk, contingency plans, and monitoring.
7. Determine need for *environmental protection measures* during construction, allocating budget and manpower for environmental supervisors to work with project engineers and architects.
8. Select *low-risk or safety-conscious designs* where this opportunity presents itself (the potential for a chemical waste disposal site to leak into an aquifer, a rail line to be located adjacent to a hospital or school zone where transport of hazardous chemical substances occurs).
9. Identify and create or enhance where feasible *amenity and disamenity landscape resources.*

---

* For a recent technical overview of the environmental impact assessment process, Rosenberg et al. (1981) provide an excellent summary; Beanlands and Duinker (1983) also provide an evaluation of the state of the art for large Northern megaprojects in Canada.

10. Identify *institutional capability* for any recommendations and recommend ways to reinforce the existing institutions, both governmental and nongovernmental. (State of the environment reports can identify such capability.)
11. Predict for future scenarios the *flexibility or reversibility* of land use design, infrastructure, and resource allocation decisions.
12. Understand *compatibilities and incompatibilities* between adjacent land uses, including pollution of all kinds, aesthetics, crop damage from vandalism or wild animals, etc.
13. Communicate *technical environmental information* in an understandable form to professionals, politicians, and the public.
14. Evaluate compliance with all *acts, regulations, and zoning*.
15. Incorporate in *official plans* environmental policies and protection requirements, compatible with existing land use and resource management policies.
16. Undertake *risk assessment* where applicable, both quantitative and qualitative—that is, the probability of a failure in a facility and the possible effects on humans and the biota.
17. Incorporate any *environmental audits* into the project design.*
18. Examine the *phasing* of a project or policy plan in terms of cost savings, social acceptability, and the advantages offered in monitoring environmental stresses at the end of each stage.
19. Develop special *environmental models* (noise, hydrology, wave energy, etc.) where appropriate.
20. Evaluate model assumptions as to applicability, relevance, and technical accuracy.

Natural Science–Based Principles

21. Understand *historical ecosystem properties* to perceive trends and to identify possible intervention strategies. (Landscape evaluation analysis is discussed in Chapter 3.)
22. Develop for future reference a systematic *inventory of existing resources* (climate, geology, soils, trace minerals, hydrology, slope, plants, animals, land use, archaeological and historic sites, and amenity resources) either qualitative or quantitative capable of being replicated or verified (Dorney 1976).
23. Develop or adapt *ecosystem models* where appropriate.
24. Predict *thresholds, lags, feedbacks, resilience, assimilative capacity, and limits* of the ecosystem based on historic trends, on an inventory of the present situation, and on a projection of the future situation.

---

* Environmental audits are industrial evaluations done by private evaluators working for but independent of the client looking for pollution compliance, complaints of an official or unofficial nature, and potential legal penalties from health or pollution damage affecting the financial viability of the industry.

25. Identify *natural processes* for land units (meso- and microclimatic scales, surface and groundwater dynamics, geochemical cycling, food chain structure, succession, erosion control, etc.), their present and potential value to humans, and cultural stresses on ecosystem processes.

26. Understand *population dynamics* of key organisms (toxic materials concentrated in food chains, tolerance and adaptation to stress, harvest potential, life strategies (*r* and *k* strategies), niche structure, ability of plants to regenerate sexually and asexually, etc.).

27. Identify *indicator plants and animal* organisms which provide insight into the quality of the environment.

28. Identify endangered species or regionally rare species.

29. Identify and control existing and new *externalities* on as small a land unit as possible, such as point and nonpoint sources of pollution; this procedure improves cyclicity, moving toward closed geochemical cycling, and reduces probability of widespread contamination in plants and animals.

30. Develop maps for the purposes of protection or enhancement of *biological productivity* of land units by understanding land capability (potential), suitability (actual productivity), and feasibility (markets, human skills, supporting infrastructure).*

31. Plan for *conservation* of nonrenewable mineral and energy resources.

32. Regulate *entropy* in ecosystems by reducing consumption and loss with design configurations and by increasing recycling or reuse of lower-grade heat sources where feasible. In this context any evolving natural system processes energy at different rates. During rapid change in a system, high levels of entropy are expected, as a change in structure requires an investment in energy, as opposed to the energy requirements in systems maintenance or preserving the steady-state situation (Jantsch 1980, pp. 52–53).

33. Identify *constraints or hazardous areas* of archaeological or historic importance; of fire potential; of a geological, pedological, hydrological, climatological, biological nature; of a toxic chemical, radiological, or technological nature, including site reuse of old industrial buildings and waste sites.

34. Identify areas serving a *landscape protection function,* such as trees growing on a steep slope providing slope stability and erosion control.

35. Identify areas offering *opportunities* for restoration, rehabilitation, enhancement, linkage, sequential use of land (e.g., mining followed by recreational use).

---

* Capability includes cropland, forestry, and wildlife as recorded by the Canada Land Inventory System; suitability measures yields as recorded by systems of the U.S. Soil Conservation Service.

36. Identify any significant *transboundary ecosystem linkages* (water, nutrient, or energy import/export).
37. Identify *unique geological and biological land units* or sensitive areas: a three-step analysis procedure of description, functions, and value is recommended (Dorney 1976). (For nature reserve designation see Balser et al. 1981.)
38. Determine ecosystem *stability-resiliency-diversity relationships* at varying scales, with awareness of the systems stability, evolution, dissipative structures paradigms (Jantsch 1980).
39. Determine *carrying capacity and assimilative capacity* limits for human population, particularly key resources such as potable water, sewage, food supply, and clean air, and for human use of landscapes such as campgrounds, parks, beaches, and lakes.
40. Identify relationship between *size* of land unit and its *biotic resources* and island *biogeography dynamics,* as size and associated edge (ecotones) appear to be important attributes of ecosystem stability and diversity.
41. Identify *health- and nuisance*-related landscape, such as pest species, zoonoses (areas where animal diseases are transmissible to humans), toxic materials, noise levels, and allergenic and toxic plants.
42. Design for, develop baseline data for, and promote *monitoring* of existing and to be built ecosystems.
43. Design *low-maintenance* landscape systems.*

Social Science–Based Principles

44. Understand *cultural linkages* between land uses, productivity, and resource recycling or reuse.
45. Identify *community and institutional values and individual concerns and perceptions* where acceptance or rejection of environmental policy, planning, and management concerns ultimately resides. Community concerns and individual concerns may or may not converge.
46. Develop *strategies* to alter human values and perceptions where this is socially and politically desirable.
47. Develop *educational* approaches at all societal levels that influence environmental perceptions and environmental abuses.
48. Identify any *values* attached to particular land areas or features such as amenity value, symbolic value, historic-archaeological value, and ethnic values in a community or individuals.
49. Map *recreational capability* at the appropriate scale (1 : 250,000, 1 : 50,000, 1 : 10,000).

---

* By way of comparison, numbers 22, 24, 25, 26, 30, and 38–41 essentially encompass the 32 principles outlined by Walker and Norton (1982).

50. Specify *public participation* approach as to the level of interaction, breadth of representation, availability of funding, and the time frame for decision making.
51. Determine *cost-benefit* relationships where applicable. Tangible and intangible aspects should not necessarily take precedence over matters of fundamental individual and community values.

Planning for ecosystem stability per se has been deliberately omitted from this list, because it is too vague a principle, leading to varying interpretations. For instance, present human densities in most countries cannot be sustained if we return to cropping stable (climax) ecosystems. Any idealistic arguments for ecosystem stability have to realize that human populations require unstable, monoculturally based agricultural ecosystems to sustain present densities. This does not diminish the need for conservation of stable or natural ecosystems serving a protective or educational role in key landscape processes. Such areas should be identified, protected, and managed to conserve the processes and educational potential explicitly identified as valuable in their own right. At a more conceptual level (Jantsch 1980), the necessity of environmental fluctuation to enhance self-organization in ecosystems suggests that planned ecosystem stability leads to eventual ecosystem collapse in the face of stress.

Diversity considerations for either a landscape mosaic or particular assemblages of plants and animals have been linked to stability and resilience, as they are related to natural science realities. Diversity-stability relationships have been considered an underlying ethic which applies to natural and social systems as outlined earlier and receive consideration in principles 21, 26, and 37.

Similarly, "island ecosystem" effects (see number 40 and Sullivan and Shaffer 1975; Burgess and Sharp 1981) are valid ecological issues as they relate to size of land units and diversity. Although this is a subject where quantitative information is becoming available, too little research has been done to date to guide decision makers except in specific situations.

Introducing both general planning principles and social and cultural principles (numbers 1–20 and 44–51) ties factors as community values to institutional capability, broadening the focus for environmental planning beyond the natural and historic resources themselves. Such linkages require that the professional try to understand the regional and local cultural mosaic prior to any planned intervention. Intervention can be sensitively tuned to local history and local aspirations, be implemented successfully, and be sustained by appropriate institutional, technical, and fiscal mechanisms if the development or policy proves satisfactory.

As regards the matter of scale, some of these principles apply more to strategic planning of large landscape areas than to urban-scale planning problems. The concept of the urban ecosystem with its zone of influence around the built city (Dorney 1983) helps identify environmental policy and zoning needs as well as urban-derived landscape evolutionary forces. Others apply to project specific planning exercises such as constructing a sewage treatment plant, a freeway, or a breakwater. There is, of course, a varying amount of overlap between these scales or situations, requiring that each must be considered on a case-by-case basis.

These principles can serve as guides when preparing environmental planning and management proposals and land use plans and when evaluating impact statements of a regional, urban, or project specific nature. For professors utilizing case studies, a checklist of these principles is useful for classroom discussions. No one discipline can organize all the information required; integrated interdisciplinary teamwork in an interactive and iterative framework is required. This reinforces the point made earlier that "environmental management" is a generic description under which individuals find a focus or foci. If the list does nothing else, it indicates the complexity of the task and the importance of bounding the environmental quality issue before undertaking studies to "answer questions." Without the principles, the studies become bottomless pits for sucking up money or for frustrating legitimate efforts of a development company or agency. These principles help to clean up what appears to many professional critics to be a messy field.

This chapter has laid the groundwork for some of the philosophical, ethical, and technical issues identified with environmental management. The organizational structure proposed by using the technical principles enables the environmental management professional to operate in a humanistic and rational mode in a world that clearly makes decisions and undertakes actions not always rationally. In addition, the merits of various philosophical viewpoints offer a foundation for discussion of issues having deep cultural and social significance.

As the field of environmental ethics develops, new insights of substantial importance will be achieved. These philosophical and ethical issues enhance the evolving sets of technical principles, whose merits can be more empirically tested in the operating real-world forum in which environmental management takes place.

Bibliography

Balser D, Bielak A, deBoer G, Tobias T, Adindu G, Dorney RS (1981) Nature reserve designation in a cultural landscape, incorporating island biogeographic theory. Lands Plan 8:329–347.

Beanlands GE, Duinker PN (1983) An Ecological Framework for Environmental Impact Assessment in Canada. Halifax: Institute for Resource and Environmental Studies, Dalhousie University.

Boughey AS (1976) Strategy for Survival. Don Mills, Ont.: Benjamin.

Brady RF, Tobias T, Eagles PFJ, Ohrner R, Micak J, Veale B, Dorney RS (1979)

A typology for an urban ecosystem and its relationship to larger biogeographical landscape units. Urban Ecol 4:11–28.

Burgess RL, Sharpe DM (1981) Forest Island Dynamics in Man-Dominated Landscapes. Ecological Studies No. 41. New York: Springer-Verlag.

Dansereau P (1973) Inscape and Landscape. Massey Lectures. Toronto: Canadian Broadcasting Company.

Dorney RS (1976) Biophysical and cultural-historic land classification and mapping for Canadian urban and urbanizing land. In, Ecological (Biophysical) Land Classification in Urban Areas. Ecol Land Class Series No. 3. Ottawa: Environment Canada.

Dorney RS (1977a) Environmental assessment: The ecological dimension. J Am Water Works Assoc 69(4):182–185.

Dorney RS (1977b) Planning for environmental quality in Canada: Perspectives for the future. Major theme paper, Can Inst of Plan Annual Meet, Toronto.

Dorney RS (1983) The urban ecosystem: Its spatial structure, its scale relationships, and its subsystem attributes. Paper presented to World Systems Conference, Caracas, Venezuela, July 10–16.

Dorney RS (1985) Predicting environmental impact of land-use development projects. In: Whitney JBR, McLaren WM (eds) Environmental Impact Assessment—The Canadian Experience. Toronto: Inst Environ Studies, pp 135–149.

Dorney RS, Hoffman DW (1979). Development of landscape planning concepts and management strategies for an urbanizing agricultural region. Lands Plan 6:151–177.

Dorney RS, Rich SG (1976) Urban design in the context of achieving environmental quality through ecosystem analysis. Contact 8(2):28–48.

Elkin T, Couture S, Palmer J (1987) State of the environment report—regional municipality of Waterloo. Working paper. University of Waterloo (Ontario): School of Urban and Regional Planning.

Findley EL (1982) The value of environmental assessment in facility development. Environ Professional 4:317–322.

Gilpin A (1980) Environmental Policy in Australia. St. Lucia: University of Queensland Press.

Holling CS (1978) Adaptive environmental assessment and management. No. 3, Intern Series on Applied Systems Analysis, International Institute for Applied Systems Analysis. Toronto: John Wiley.

Holling CS, Goldberg MA (1971) Ecology and planning. J Am Inst Plan 37:221–230.

Jantsch E (1980) The Self-Organizing Universe. New York: Pergamon Press.

Kaiser EJ et al. (1974) Promoting environmental quality through urban planning and controls. EPA-600/5-73-015. Washington: U.S. Government Printing Office.

Loucks DE, Perkowski J, Bowie DB (1982) The impact of environmental assessment on energy project development. Calgary, Alta.: Petro-Canada.

Mattyasovszky E (1975) Key principles in planning for environmental quality. Plan Can, 38–43. Vol. 15/1 Mr.

McHarg I (1969) Design with Nature. Garden City, NY: Natural History Press.

Meadows DL et al. (1972) Limits to Growth. New York: Universe Books.

Odum EP (1969) The strategy of ecosystem development. Science 164:262–270.

Ontario Society for Environmental Management (1976) Constitution. Waterloo: University of Waterloo Press.

Reynolds PD (1981) Constraints on the ethical activities of professional societies. Paper presented at Am Assoc Adv Sci Mtg, Toronto.

Rosenberg DM et al. (1981) Recent trends in environmental impact assessment. Can J Fish Aquatic Sci 38:591–624.

Shrader-Frechette KS (1982) Environmental impact assessment and the fallacy of unfinished business. Environ Ethics 4(1):37–47.

Sullivan AL, Shaffer ML (1975) Biogeography of the megazoo. Science 189:13–17.

U.S. Central Intelligence Agency (1976) A study of climatological research as it pertains to intelligence problems. Washington: CIA.

Walker BH, Norton GA (1982) Applied ecology: Towards a positive approach. II. Applied ecological analysis J Environ Mgmt 14:325–342.

3
The Conceptual Basis for Environmental Management

The conceptual basis for environmental management deserves careful consideration. Without first developing a common conceptual basis, the professional area of environmental management appears to be on shaky ground and less a science than an art. Issues such as the appropriate landscape classification, the typology to be used, the appropriate scale of mapping, and the identification of the mechanisms driving landscape evolution, unless clarified, lead to apparent, more than real, differences as perceived by outsiders. If conceptual agreement proves less than adequate, raising it here as an issue serves as a beginning in understanding the confrontations that develop between lawyers in front of a hearing board, the hard questions asked by a client or by politicians, or technical disagreement between environmental managers and allied professionals.

Depending on the terms of reference initially laid down for the study, an analysis judged ''inappropriate'' may require the environmental manager to undertake additional work. Often this is done at his or her own expense, including reorganizing the maps and text, new fieldwork, and more meetings with agencies to negotiate salient issues. In addition, an inappropriate or incomplete environmental analysis, stemming from a poor conceptual framework, spawns a host of problems in dealing with the media and the public. Once public or legal hearings commence, it is usually too late to redefine the problem or to undertake additional fieldwork. An image tarnished in the arena of public or legal debate requires time to repolish, if ever. As is true in the legal and medical professions, failures always attract greater publicity than day-to-day excellence.

Legal damages or a liability suit for work badly performed becomes a specter as well. Poorly conceived work and fieldwork poorly executed result in extensive losses to the client. These losses may be the result of delays because hearings are recessed and then rescheduled, or they may be due to cost overruns during facility construction because of incorrect biophysical mapping. Such cost overruns result from higher front-end costs, such as interest charges, penalties allowed in contracts for delay of construction, shutdowns, "extras" for the contractor due to special labor or equipment requirements, or new engineering design work.

This chapter begins with definitions of environment and ecosystem and then moves on to questions of landscape classification, followed by examples from each of the modes presented in Fig. 1-1.

Definition of Environment

One of the initial conceptual questions that must be answered is what is included or excluded in the term "environment." The boundary placed on the definition by some biological scientists, restricting the definition to include only abiotic and biotic interactions (biophysical ecosystems and interactions), stems from the belief that human cultural influences are too complex and confusing. They should be studied separately from the natural processes. Some environmental assessment procedures (like those of the Canadian federal government) follow this biophysical mode, while others (like those in the United States and Ontario) include human influences in their operational definitions.

My experience as a practitioner demonstrates that the impact of a proposed facility and the effect of a proposed environmental policy in an official plan sooner or later focus on socioeconomic and related health concerns in evidence presented before a hearing body. The pure biophysical information has reduced meaning when isolated from human concerns. For these reasons, then, the only acceptable definition of environment on pragmatic grounds must include human institutions and activities as they influence and are influenced by biophysical processes.

Thus abiotic–biotic–cultural (ABC) definition of environment is followed in this book. Using this definition of environment leads to a comprehensive discussion of ecosystems, landscape classification and scale typologies, and identification of the importance of landscape evolution. This discussion is followed by an examination of environmental planning and environmental protection paradigms in this human nature context.

Definition of Ecosystem

If environment is seen in an ABC context, ecosystems are discrete units of the environment. Specifically, they can be defined as being open systems, having identifiable boundaries and encompassing both structural components, system properties, and processes. An open system means a system having material and energy flow across its boundaries. Systems properties include such aspects as thresholds, lags, and feedback, while processes include energy flux, geochemical cycling, and food chain accumulation of toxic materials.*

Landscape Classification and Scale

The many ways landscape is visualized "in the eye of the beholder." Some beholders have a romantic vision, others an historic one, while others may see production, economics, ecological processes, patriotism, or even colonialism in the landscape mosaic. The environmental manager should be aware that multiple visions are to be expected. Identifying quantitative classification systems, which are agreeable to all parties, serves as a point of departure for budgeting, composition of the interdisciplinary team, and time constraints (if seasonal phenomena are to be included). However, the critical question is what questions need to be answered, since the methodologies and scales vary in their ability to communicate answers to particular and specific questions.

Although the landscape classification systems contrasted here are hardly exhaustive, the broad outlines of scientific classification systems should become clearer in Table 3-1. These selected classification systems are categorized into four general groups: descriptive, ecological, features, and derivative. Three of these four groups, the ecological group excepted, describe the environment in a static context; in addition most are constrained by one or a few scales. Dynamic processes, if deemed important, may need to be identified and described in verbal or schematic ways.

A somewhat simpler landscape classification system is proposed in Fig. 3-1, based on three ecosystem types—natural, agricultural, and urban. The urban ecosystem, divided into three zones (built city, urban fringe, and urban shadow) includes all land within the daily commuting radius of a city (25,000 population). The urban fringe is the area next to the built city and is slated for early development. Natural ecosystems and agroecosystems have a high percentage of land in natural vegetation and include managed forest lands and nature reserves. The advantage of this classification is its simplicity and adaptability to variable scales. It can serve as an initial entry point to help in defining the scope of study and

---

* How to deal with extraterrestrial space poses an interesting, and no longer academic, question. Given the assembling of spacecraft and possible moon colonization, I believe it is useful to include these areas of an ecosystem in our definition, since they fit all the criteria outlined above.

Table 3-1. Some comparative landscape classification approaches.

	Time frame	Approximate scale	References
Descriptive			
Land use	Past, present	1:50,000 to 1:25,000	Gierman (1981)
Zoning maps	Past, present	1:5,000	General use
Ecological			
Abiotic–biotic spatial configurations	Present	1:250,000 to 1:50,000	Hills (1974), Cassie et al. (1970), Moss (1983)
Trophic structure	Present	1:25,000	Dansereau and Pare (1977)
Intrinsic suitability	Present	1:25,000	McHarg (1969)
Ecosystems	Present	1:3,000,000 to 1:2,500	This book (Fig. 3-1)
Landscape processes	Present	1:250,000 to 1:2,500	Moss (1984)
Features			
ABC (abiotic–biotic–cultural)	Present	1:250,000 to 1:25,000	Dorney (1977)
Metland	Present, future	1:25,000	Fabos and Caswell (1977)
Point mapping (Geomorphology, soils, land use)	Present, future	1:20,000	Coleman (1975)
Urban trees	Present	1:500	Dorney et al. (1983)
Derivitive			
Capability maps	Present to future potential production	1:250,000	Canada Land Inventory (1969–1977)
Symapping	Present, future	1:20,000	Coleman and MacNaughton (1971)
Landsat (electromagnetic) imagery	Present	1:1,000,000 to 1:50,000	van Claasan and Ross (1981)

LANDSCAPE TYPES

Scale	Approx. Map Scale	Minimum Mappable Unit Size	Natural Ecosystem**		Agroecosystem***		Urban Ecosystem****		Examples of Urban Ecosystems
			Agroecosystem Island	Urban Ecosystem Island	Natural Ecosystem Island	Urban Ecosystem Island	Agroecosystem Island	Natural Ecosystem Island	
Ecoregion Scale	1 : 1,000,000 - 1 : 3,000,000	15km². - 150km²	Agroecosystem Island	Urban Ecosystem Island	Natural Ecosystem Island	Urban Ecosystem Island	Agroecosystem Island	Natural Ecosystem Island	Toronto-Windsor corridor (Ont.)
Ecodistrict Scale	1 : 125,000 - 1 : 500,000	160- 400 ha	Agroecosystem Island	Urban Ecosystem Island	Natural Ecosystem Island	Urban Ecosystem Island	Agroecosystem Island	Natural Ecosystem Island	Metro Toronto (Ont.)
Ecosection Scale	1 : 50,000 - 1 : 250,000	6-60 ha	Agroecosystem Island	Urban Ecosystem Island	Natural Ecosystem Island	Urban Ecosystem Island	Agroecosystem Island	Natural Ecosystem Island	London (Ont.)
Ecosite Scale	1 : 10,000 - 1 : 50,000	1-4 ha	Agroecosystem Island	Urban Ecosystem Island	Natural Ecosystem Island	Urban Ecosystem Island	Agroecosystem Island	Natural Ecosystem Island	Waterloo (Ont.)
Ecoelement Scale	1 : 2,500	.01- .02 ha							None*****

* Adapted from Baker et al., 1981.
** More than 50% natural vegetation cover, managed or unmanaged.
*** More than 50% of land surface in agricultural production.
**** More than 50% of land surface within influence of three urban zones (built city, urban fringe, urban shadow).
***** Area is too small to be a city. Hamlets would be mapped at this scale.

Figure 3-1. A landscape classification based on three ecosystem types. *

then to be followed by one or more detailed mapping procedures. Agro- and urban ecosystems can be combined into a category called "cultural ecosystems," although the separation proposed in Fig. 3-1 seems to make practical sense. The salient point is that the complexity of urban ecosystems requires considerably more analysis of the human factors than does the natural.

If a classification system based on static features is selected, as highlighted in Table 3-1, it should not be assumed that ecological processes may be overlooked in subsequent hearings or project review. Although the environmental manager may map or describe the environment in fairly static terms, ecological linkages and processes can be shown through graphic and tabular means. Future dynamic scenarios can be easily generated, evaluated, and displayed by computer simulation, as in Metland, Point Mapping, and Symapping (see Table 3-1).

The environmental manager has numerous ways to conceptualize and to depict land use or landscape mosaics. Some systems (Table 3-1) are more comprehensive, while others are more graphic or rely more on tabular data. Others, for example Dansereau's trophic structure maps oriented to ecological processes, are expensive to do and somewhat inflexible as to scale and time. Some, such as the ABC feature maps and the capability maps for crop production, are more readily understood by the public than are computer-printed maps such as symaps or the ecological spatial configuration maps of Hills (1974) and Moss (1983). Whatever mapping system is selected, its ability to depict the problems identified, its accuracy, and its ease in communication to the target audience are paramount considerations for the environmental manager to consider before beginning the work, and in fact as early on as preparing the proposal and budget.

Landscape Evolution

Accepting the fact that both agro- and urban ecosystems have landscape mosaics that change over time because of human activity, two conceptual points are pertinent. First is to determine the underlying human factors or stresses that cause these changes. Next is justifying the interventions by which the environmental manager guides the process instead of responding to changes imposed by factors and actors outside the frame of reference.

Those major factors, modified from an earlier version (Dorney 1985), which appear to control and alter the natural and cultural landscape mosaic over time, are five: institutional, social, technological, economic, and ecological (Fig. 3-2). The interplay of these factors is complex in time and space, but if each factor is seen as a stress predominating at varying points in time to be replaced by another factor or combination of factors

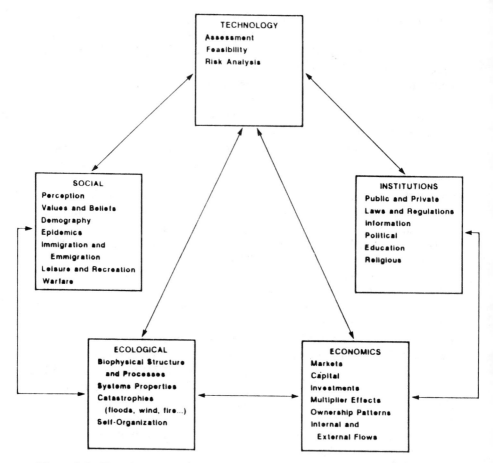

Figure 3-2. Five factors or driving mechanisms responsible for the landscape mosaic over time. Note: In the natural ecosystem the "ecological factor" would be the prime mechanism, while in the agroecosystem and urban ecosystem this factor would be subordinate to the other four factors.

predominating later, then the past and present landscapes have various alternative trajectories (Fig. 3-3). This means that various end states at any one time can be postulated; four are shown, lettered A (the baseline) to A3.

An analysis of the various specific stresses for southern Ontario over a 150-year time frame shows that of the many stresses causing change, few will be repeated again. For example, in Fig. 3-3 the initial social factor was the European settlers' image of how a settled landscape should look shown as t1; cleared land along the roads and woodlots on back of farm (ca. 1800–1850). The economic factors of an export-based colonial agriculture determined the structure of markets (1830–1900) shown as t2 evidenced in the grid patterns of the agricultural landscape. The technol-

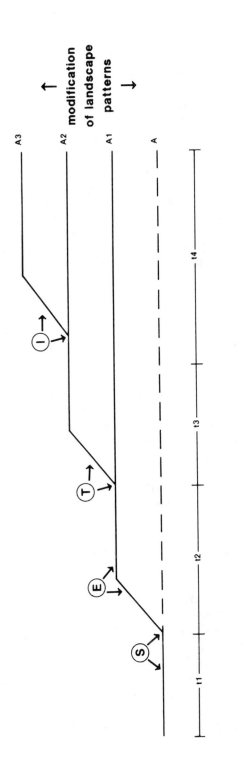

Figure 3-3. Landscape evolution based on varying stresses and decision points. The initial trajectory (dashed line) is the baseline where no stresses or interventions occurred.

ogy, identified as T in t3, is the railroads. In the 1870s the location of railroads altered urban growth and industrial development, while tourism was minimal. In the future these factors or stressors will be modified if tourism represented by the institutional block in t4 is promoted actively.

The use of technology assessment tools coupled to Delphi forecasting techniques allows for landscape features or scenarios that may be affected by new technologies, new economic forces, new social dynamics, and new institutional dynamics. It then becomes appropriate for the environmental manager, as part of a team effort, to identify these stresses and to suggest interventions to control, to redirect, and to accelerate the likely situations. Thus, changes that are environmentally positive can be encouraged, with appropriate controls developed for those that are of a neutral or a negative nature. The definition of "positive" is complex and somewhat "in the eye of the beholder": the technical and philosophical issues discussed in Chapter 2 bear directly on how to analyze the environmental context of such issues so that holistic analyses can be proposed.

The concept of cultural landscape evolution is a useful planning tool, because it focuses the discussion about past, present, and future stresses. Discussion translates into scenarios, some of which could lead to appropriate or inappropriate future land use. Positive planning occurs, as opposed to reactive ad hoc planning. Furthermore, the terms "conservation," "preservation," and "rehabilitation" can be cast in a continuum of change with human perceptions, institutional adequacy, and economics as a focal point for making appropriate decisions. Forces within and outside the control of land managers are recognized so that the limited effort available is focused on critical points affecting the landscape evolutionary trajectory rather than on waiting until the trajectory has begun and then trying to find funds to modify or mitigate the effects or to develop the political will to stop change altogether.

Planners are continually attempting models to optimize land use and plan the future countryside (see B.H. Green in Roberts and Roberts 1984, p. 205). Consistently, the impact of new technology is not included in the causal mechanisms altering land use change; an exception is the conceptual framework for environmental management outlined by Garlauskas (1975). Economics, social and ecological forces, and institutional mechanisms are the four factors commonly identified. Adding technology and technology assessment techniques to a generalized model for future land use provides a greater degree of realism and a higher level of predictability, removing some of the surprise. For example, if biotechnology rapidly increases agricultural productivity, large amounts of agricultural land will become surplus. Future uses of this acreage could be planned, or market forces could make the decision. Further, the socioeconomic and ecological impact would differ widely depending on how and where the land retirement takes place.

Specific examples of conceptual issues likely encountered in each of the two phases and six modes in Fig. 1-1 will now be presented. Although not comprehensive in any sense, examples from both aquatic and terrestrial sources should help to clarify the discussion.

Ecoplanning Phase

Planning as a sequence of steps, or a "planning process," has evolved from the professions of architecture and civil engineering. Initially, this process was devoid of anything but rudimentary biophysical data. Conceptually, this process occurs for various kinds of urban or park planning—the concept plan, the master plan, the secondary plan, and the detailed design plan (Fig. 3-4). These plans form a sequence of increasing detail (left to right) and an associated increase in level of costs.

The conceptual coupling of specific biophysical information (Fig. 3-4, lower portion) to the sequence of planning detail is important to appreciate. The biophysical information generated must be targeted to a specific kind of plan to be useful to other professionals and to be cost-effective as well. This conceptualization of a holistic planning process leads from the ecoplanning phase to be followed by an environmental protection phase (Fig. 1-1) to ensure correct implementation.

The six modes of ecoplanning are the first part of an environmental management process (Fig. 1-1). Ecoplanning initiates understanding on how an acceptable human nature fit can be made at both the urban and regional level of detail for both aquatic and terrestrial ecosystems. Although any landscape analysis or classification system can be used for ecoplanning, because of simplicity in mapping, ease in public presentations, and adaptability to any scale, I prefer the ABC method (Table 3-1).

Ecoplanning, moving from left to right (Fig. 1-1), proceeds through various stages. Combined with the technical principles (see Chapter 2), these sequential stages, as well as defining the six modes (shown as vertical sections in Fig. 1-1), do a number of other things that former environmental planning approaches do not, or do not clearly articulate. Ecoplanning does the following:

1. Probes questions, which may not be amenable to quantitative analysis, about *historic ecosystem structure* so that current patterns of natural resources utilization and present and future structure and function of the ecosystem can be better understood.
2. Enlarges upon the mapping of *present* natural resource patterns of land and associated shoreline uses (generally the main effort in any environmental analysis) by exploring the *potential or future* landscape evolution. This may result in considerable ecosystem modification and restoration.

STAGE	CONCEPT	MASTER PLAN	SECONDARY PLAN	DETAILED DESIGN	IMPLEMENTATION and MONITORING
TYPE OF PLAN	Strategic	General or "Official"	Structure	Subdivision or Site	Project Management
CONVENTIONAL CONSIDERATIONS	Studies of Need and Feasibility	Choice of Options	Large Scale Design	Detailed Design	Supervision
ENVIRONMENTAL CONSIDERATIONS	Land Suitability and Capability	Appraisal of Impacts, Opportunities and Constraints	Environmental Assessment and Ecosystem Modification		Resource Management and Environmental Supervision
ENVIRONMENTAL DATA BASE	Existing Records and Reconnaissance Surveys	Increasing Detail Appropriate to the Stage in the Process, Including Field Inventories of Soils, Vegetation, Ground Water, etc. Also Data Based on Consultation with Local, Regional and Provincial Authorities.			Maintaining and Response to Effects of Implementation
REPORTS		At Increasing Levels of Detail General or Technical Language, Depending on Purpose of Document.			Progress Reports and Remedial Recommendations
MAP SCALE	1 : 250,000 1 : 50,000 1 : 25,000	1 : 50,000 1 : 25,000 1 : 12,000	1 : 10,000 1 : 5,000	1 : 5,000 1 : 1,200 1 : 500	

Figure 3-4. Conventional planning process for urban development from the conceptual stage to the detailed design stage. Courtesy of Ecoplans Ltd.

3. Enumerates likely or possible environmental gains resulting from any course of action; it is not merely a list of environmental losses.
4. Places these issues of historic dynamics and losses or gains into a context of environmental protection to ensure that the environmental quality envisioned in the ecoplanning phase is in fact delivered and satisfies all appropriate regulations and expectations.
5. Includes explicitly the issue of interprofessional, professional, and public communication.

In this context, ecoplanning is not another new environmental planning methodology but a comprehensive procedure for developing an ideal environmental information and management system which builds on existing and well-established planning processes. It can be adapted to planning of aquatic and terrestrial ecosystems ranging from the regional scale to 1 square yard or 1 square meter to the scale of a domestic garden, and perhaps at the larger continental or macroregional scale as well.

In Fig. 1-1 the schematic outline of the environmental management process shows the field split into two phases. This first phase is ecoplanning essentially a "paper exercise," which is followed by the second or "hard-hat" protection phase, which is the implementation aspect. Further amplification of Fig. 1-1 is warranted to expand on the conceptual environmental management aspects and to explain in more detail what environmental management does and how it can accomplish its goals. This is depicted in Fig. 3-5.

Some stages (Fig. 3-5) are common to modes grouped into two blocks (the horizontal components). Each of these stages may have common aspects that can be illustrated at this point for one mode, the new-facility development mode (Table 3-2), thus clarifying the level of detail required. These are the terms of reference, followed by a study design, a justification, and in some cases a policy aspect. The other stages—evaluation and design, and hearings and approvals—are reasonably self-evident.

In the analytical stage, the next stage, certain common elements for a new facility also appear and can be elaborated on at this point (Table 3-3). Other elements, such as risk information storage and retrieval and risk analysis, are self-explanatory. Other authors who have attempted a similar list of elements can be consulted for comparative purposes (Roberts and Roberts 1984, p. 21).

Having identified many of the common vertical elements in Fig. 3-5, discussion of the six horizontal modes in this figure will clarify the ecoplanning phase, or paper exercise, of environmental management.

Urban and Regional Development Mode

This mode relates to the production of a development plan and associated implementation efforts. It is usually done by government and designed to

Ecoplanning Phase

	COORDINATION AND ADMINISTRATION STAGE	ANALYTICAL STAGE	EVALUATION AND DESIGN STAGE		REPORT STAGE	HEARINGS AND APPROVALS STAGE	
	Project Administration	Environmental Analysis	Environmental Synthesis	Environmental Design	Public Summary Reports / Environmental Impact Statement	Public Hearings and Approval / Environmental Pre-Hearing · Environmental Mediation	Implementing Legislation and Funding / Environmental Hearing and Approval
URBAN AND REGIONAL DEVELOPMENT MODE	-terms of reference -study design -justification -policy	-historical landscape dynamics -technical principles -description of existing environment -landscape functions -landscape values	-historical processes -present processes -future processes -landscape evolution	-open space -production landscapes -urban development	Public Summary Reports -technical documents	Public Hearings and Approval	Implementing Legislation and Funding
NEW FACILITY DEVELOPMENT MODE	-terms of reference -study design -justification -policy analysis	-technical principles -description of existing environment -landscape functions -landscape values		-siting -aesthetics	Environmental Impact Statement -alternatives -abiotic, biotic & cultural impacts -mitigation -rehabilitation -contingencies	Environmental Pre-Hearing Environmental Mediation	Environmental Hearing and Approval -with conditions -without conditions
OLD FACILITY DECOMMISSIONING MODE	-terms of reference -study design -justification	-technical principles -description of existing environment -landscape functions -landscape values		-siting -aesthetics	Environmental Impact Statement -alternatives -risk analysis -emergency planning -mitigation -rehabilitation		Environmental Hearing and Approval -with conditions -without conditions

	COORDINATION AND ADMINISTRATION STAGE	COMPLIANCE SCREENING STAGE	EVALUATION AND ASSESSMENT STAGE	PUBLIC PARTICIPATION STAGE	REPORT OR COMPLIANCE AUDIT STAGE	HEARING AND APPROVAL STAGE	REGULATION FORMULATION AND DISSEMINATION STAGE
GOVERNMENT POLICY FORMULATION MODE	Interdepartmental Coordinating Committee or Policy Secretariat; -terms of reference	Environmental Analysis; -socio-economic data; -biophysical data; -enforcement data; -ecosystem stress data	Policy Papers Issued as Needed	Briefs From Public Solicited	State of the Environment Reports	as Required by Law	Proposing of -new laws; -new regulations; -new by-laws; -new planning processes
URBAN AND REGIONAL GOVERNMENT OPERATIONAL MODE	Environmental Quality Secretariat or Coordinator; -goals; -objectives; -policies	Compilation Data on Local-Regional Concerns, Prosecutions, Permits, Licenses, Future Issues Analysis	Inter-Agency Discussion Standards Review	Key Groups and General Public Meetings as Needed	Reports on Specific Topics	Legal or Quasi-Legal Approvals Required	Issuance By-Laws, Regulations, and Operational Handbook
CORPORATE FACILITY OPERATION MODE	Environmental Compliance Section; -terms of reference; -reporting procedures	Analysis of In-House Product Streams and Waste Generation-Disposal (cradle to grave analysis)	Environmental Compliance Audit Government Dialogue	as Needed	Environmental Audit Reports	as Required by Government	In-House Only

Figure 3-5. The detailed environmental management process by phases, modes, and stages.

Environmental Protection Phase

	CONTRACT STAGE	CONSTRUCTION STAGE	REPORT STAGE	ABATEMENT AND ENFORCEMENT STAGE	REHABILITATION STAGE	MONITORING STAGE	RESEARCH STAGE
URBAN AND REGIONAL DEVELOPMENT MODE			not applicable				
NEW FACILITY DEVELOPMENT MODE	Contract Specification Writing -sensitive area protection -mitigation -rehabilitation	Pre-Construction Environmental Assessment Summary Construction Supervision for Environmental Concerns	as Built Report	Liason Required as Part of Regional Operation Mode	Restoration and Rehabilitation -terrestrial -aquatic	Monitoring -terrestrial -aquatic -bio-indicators	as Needed
OLD FACILITY DEVELOPMENT MODE	Contract Specification Writing -sensitive area protection	Pre-Demolition Environmental Assessment Summary Demolition Supervision for Environmental Concerns	as Demolished Report	Liason Required as Part of Regional Operation Mode	Rehabilitation -terrestrial -aquatic	Monitoring -terrestrial -aquatic -bio-indicators	Special Contracts to Facilitate Short-Cuts

Environmental Protection Phase (con't)

	COORDINATION AND ADMINISTRATION STAGE	COMPLIANCE SCREENING STAGE	EVALUATION AND ASSESSMENT STAGE	PUBLIC PARTICIPATION STAGE	REPORT ON COMPLIANCE AUDIT STAGE	HEARING AND APPROVAL STAGE	REGULATION FORMULATION AND DISSEMINATION STAGE
GOVERNMENT POLICY FORMULATION MODE	not applicable		Special Studies as Required	not applicable	Special Reports as Needed	not applicable	
URBAN AND REGIONAL GOVERNMENT OPERATION MODE	Laboratory and Field Staff Organized with Terms of Reference	Monitoring and Field Enforcement/ Inspection on a Regular Basis for Infractions and Compliance	Reports and Meetings as Required	not applicable	Stop-Work Orders, Approval of as Built Reports and Recommendations for Prosecution	not applicable	Summary Reports as Required
CORPORATE FACILITY OPERATION MODE	In-House Personnel or Consultants Given Terms of Reference	Periodic Analysis as Required of Wastes	Periodic Studies to Become Part of Environmental Audit	not applicable	Periodic Reports to Become Part of Environmental Audit	not applicable	In-House Procedures Established Only

Figure 3-5. The detailed environmental management process by phases, modes, and stages.

Table 3-2. Elements to be analyzed by client and consultant as part of the organization and administration stage of a new facility development (such as a power plant).[a]

Elements	Specific concerns
Terms of reference	Write clear statement identifying interaction between the problem, the time, and cost to do the work
Study design	Undertake critical path analysis (a matrix of time and apsects to be studied)
	Locate required specialists who can work within a critical path framework
	Develop budget and sign separate contracts with the specialists
	Design information requirements to ensure readability, the target by audience(s), and credibility of technical information
	Design a graphic system (scale, color, sketches, photographs)
Justification	Clarify client's rationales for the project so an adequate case for the action can or cannot be established on technical grounds. Should the project prove not to be justified on technical grounds, the client should be aware that the environmental analysis may jeopardize the project approval or alter the project
Policy	Identify existing policies that may affect the scope of the environmental work: explicit policies, bylaws, laws in force, and less specific and often unwritten administrative guidelines

[a] See far left vertical column of Fig. 3-5.

accelerate investment in a defined geographic urban or regional area. Examples of this mode are county recreation and tourism studies, such as Montgomery County, New York; the Niagara Escarpment Plan in Ontario in the 1970s; a four-nation river basin development scheme such as the Rio de la Plata project in South America and the Potomac River basin plan in Washington, D.C. (Potomac Planning Task Force 1967) in the 1960s; an urban waterfront redevelopment in Toronto in the 1980s (Dorney 1986); a new resource town and mine in the north such as Pickle Lake, Ontario, in the 1970s; and Scottish oil and gas development (Currie and MacLellan in Roberts and Roberts 1984, p. 23). These studies generally have a high public profile, goals are passed to the development team from the top down, political backing or justification is evident, and there may be many policy implications left hanging for later resolution.

If an environmental analysis stage is included explicitly in this mode (see Fig. 3-4), it follows from the terms of reference developed in the project administration stage. Subsequent stages as part of the overall planning process include an analytical stage, an evaluation and design stage, and finally a public hearings and approval stage combined with implementing legislation and funding.

The high profile and rapid implementation of these development modes overshadows, in terms of environmental impact, by far any single facility development that comes if and when the urban or regional development proceeds. If environmental quality concerns are not considered initially in the development mode, the subsequent impact statements on separate facilities become ad hoc planning approaches, perhaps mitigating damage, but not altering the larger picture.

Examples in Ontario where environmental quality concerns were integrated into such regional development modes are the Nanticoke Industrial Basin along Lake Erie's north shore. From initial announcements in the late 1960s of a new steel mill to be located on the shoreline, regional environmental planning studies were initiated by the government (Chanasyk 1970) followed by, among others, an environmental planning study for the new town of Townsend (Dorney 1977). At the same time, baseline and pollution monitoring studies were initiated to protect the coastal environment under the jurisdiction of a committee funded by both government and major industries. Calefaction from discharge cooling water, entrainment of small fish on water intake screens, gaseous emissions, shoreline structures and their effect on fish migration, and shoreline erosion were some of the issues pinpointed. Such up-front environmental initiatives in regional development planning were not common prior to 1970 but today are becoming worldwide. Shore management guides are now available, for example, for the Great Lakes (Fisheries and Oceans 1981).

The justification for inclusion of environmental quality concerns into such regional development projects is based on economic efficiency, on control of externalities, and on the broader questions of social justice. Mistakes made in ignorance, however well-meaning the project appears to its backers, cause escalating project costs to mitigate damage and can produce an economic and social white elephant or even a political backlash. The South Indian Lake Water diversion study for Manitoba (Canada) Hydro done in the early 1970s had such a backlash. The water diversion for an energy scheme caused a sufficient public outcry to unseat the New Democratic Party government. The issue was the social impact of the water development on various Indian bands. Ironically, in spite of the political change in mandate, the new government built a reduced hydroelectric scheme. Today the externalities from these projects (shoreline erosion, trash in fishing nets) are resulting in substantial and

Table 3-3. Common elements to be analyzed by client and consultant as part of the analytical stage for a new facility development.[a]

Elements	Specific concerns
Historical landscape dynamics	Identify: 1. Geological factors affecting land and seashore form 2. Pre-Caucasian floristic and faunistic composition related to physiographic, climatic features, ecotones 3. Trends in floristic, faunistic, land use, and/or marine diversity 4. Trends in land use for crops, types, and numbers of livestock (by decade or half-century) and shoreline uses 5. Impact of technology on agricultural production and land use over time (field size, demise of terraces, use of fertilizers, pesticides, herbicides) or in natural landscapes on wildlife and natural plant cover (such as overhunting, forest fires) or on marine resources 6. Ethnic-cultural factors affecting landowners' perception, use, or enjoyment of the land surface or marine resources 7. Changing markets for agricultural, fish, game, and forest products
Technical principles	For list of principles see Chapter 2
Description of existing environment	Describe: 1. Abiotic elements (topography, climate, noise, water, geology, soils), wave energy, shoreline erosion 2. Biotic elements (plants, animals) 3. Cultural aspects (historic structures, archaeological sites, cemeteries, landscape aesthetics)
Landscape functions	Identify biophysical processes benefiting or harming humans: 1. Hydrology (qualitative and quantitative measurements) as affected by land uses 2. Groundwater recharge zones and its subsurface movement 3. Soil capability for various common and specialized field and orchard crops, forest timber, and pulpwood production

4. Pollution abatement functions of natural systems (such as a marsh reducing phosphorus content in water)
5. Containment patterns:
   a. windbreaks reducing blowing sand, preventing crop damage, and modifying microclimate for human comfort
   b. noise buffers
   c. visual buffers
   d. erosion control by vegetation
   e. flood control works
   f. sand dunes controlling movement and damage
6. Forested valley lands delivering cool-filtered air to an urban heat island
7. Reduced costs of natural, as opposed to managed, plant communities
8. Hazardous lands (slippage areas, fault zones, organic soils, natural gas seepage, hurricane zones, volcanic zones, etc.)
9. Medically hazardous or toxic areas to humans and animals
10. Epidemiologically or epizootiologically hazardous zones
11. Cropland or forest diseases related to pest populations, predatory insects, hosts for rusts

Landscape values   Identify:
1. Human perceptions attaching value to objects or processes:
   a. Unique historical resources (single artifacts and groups of artifacts)
   b. Unique archaeological resources
   c. Unique natural areas (geologically or biologically significant)
2. Land capability (potential without respect to human management) for recreation
3. Landscape aesthetics (viewpoints, color, variety, form)
4. Hunting, fishing, wildlife viewing, and collecting areas
5. Rare and endangered flora and fauna

---

[a] See second left vertical column in Fig. 3-5.

embarrassing out-of-court settlements by Manitoba Hydro to native fishermen (D. Witty, personal communication).

In the 1970s the United States Fish and Wildlife Service contracted coastal biological inventory mapping (Beccasio et al. 1980). This came about because regional development and new facility development were affecting sensitive estuaries and critical coastlines. Without a comprehensive data base on spawning and nesting areas, winter concentration areas for migratory birds, etc., it was tedious for service personnel to respond to inquiries or to contest data provided by various proponents who wished to dredge or build on the land-water interface. The map series, now complete for the East Coast of the United States, have been useful, although new information and natural changes since the mapping was done require some modification of the earlier map series.

New-Facility Mode

The development of a new facility or a construction project such as a pipeline, highway, subdivision, or hospital follows an approval sequence similar to the Urban and Regional Development Mode, except that for some projects an evironmental impact statement (EIS) or an environmental assessment report (EAR) is prepared. For a subdivision plan, a less formal environmental report (ER) may be prepared prior to hearings and approvals (Fig. 3-6). Usually included in this EIS, EAR, or ER are procedures for mitigation, for rehabilitation contingencies, and for monitoring as part of the environmental protection phase. An example of a mitigating procedure is a recommendation for using heavy equipment on sensitive cropland only when the ground is frozen to a depth of 16 inches or 40 cm, or for crossing a trout stream in summer wth a pipeline after the brook trout have bred in fall and the eggs have hatched in spring.

Among the large body of literature on environmental assessment, some good summaries are available. One that contrasts current EIA procedures in Western countries and provides a useful overview is in Chapter 4 of Roberts and Roberts (1984). In this same volume, W.R. Effer (p. 283) has documented in detail procedures of a hydroelectric authority in selecting a new generating site.

Contingencies refer to what do we do if a certain situation occurs. For example, if a gasoline product line breaks or two oil tankers collide, what spill and fire control procedures are available? Comments on contingencies should be limited to those within the control of the client. For example, commenting in a report on the possibility of a highway collision of a liquid propane tank truck may be outside the legal mandate of the hearing body debating a truck terminal if another hearing body regulates highway safety.

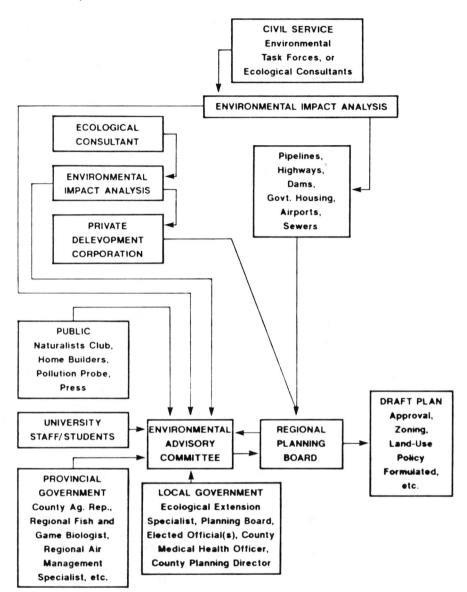

Figure 3-6. An organizational framework for assessing the environmental impact of facility development at the regional planning level.

The aspects of study design (Fig. 3-5) refer to the location of constructs in relationship to biophysical features and ecological processes. For example, a sewer line passing near a woodland growing on poorly drained soils usually lowers the water table, unless special cement collars are installed along the line. The impact on the stand is beneficial if the tree species (e.g., white ash, hard maple) are both tolerant of drier conditions and young in age (less than 35 years); the impact is disastrous if the stand is hemlock and yellow birch of a mature age (over 75 years). The effect depends on what is desired and then ensuring that the line is designed accordingly. In Chapter 1, four levels of environmental planning and design were identified—flat-earth planning, contour planning, feature planning, and ecoplanning (see also Dorney and Rich 1976). These levels reinforce the importance of design in achieving a comfortable human/nature fit.

For oceanic developments, the force of hurricane winds must be considered in the design of breakwaters, seawalls, and buildings fronting on the ocean. Shoreline stability, ocean and Great Lakes water levels, littoral drift, and ice damage on rock piers constrain land-water interface design (Dolan and Lins 1987; Pikey et al. 1983) and regulatory bodies' response. In estuarine situations, ecosystem models are useful (Nemerow 1985) in dealing with pollution loadings.

Environmental mediation, like any mediation, can be less costly than a hearing, especially a contentious hearing. To be successful, all the actors must be identified and have the opportunity to participate if they wish to do so. Project proponents are often prepared to undertake expensive mitigation or rehabilitation, especially if the alternative is a long-delayed hearing. Site-specific arguments tend to be more mediable than rights-of-way arguments (Dorney and Smith 1985).

If an EIS or EAR is not legally required, a similar but less formal ecoplanning and environmental protection process can be followed by the planning authority. As I have seen over and over again, once legal prescriptions are relaxed, the environmental planning, protection, and inspection become sloppy.

Old-Facility Decommissioning Mode

Little experience is available on procedures that are environmentally sound for decommissioning facilities. The Three Mile Island Nuclear plant in Pennsylvania may be, for example, one of the first large energy facilities to be decommissioned. The general planning process followed in the New Facility Development Mode is applicable, except in reverse—ending up with a green field where once a facility stood. The critical aspect, however, is toxicants and decontamination.

In a recent exploratory study for a city in south-central Ontario (McLean 1986), 72 potentially hazardous sites and reuse locations were identified from old insurance records. These sites, in the old industrial core of the city, were factory sites used for gas production from coal, sawmills, railway yards, and chemical depots. All of these land uses contain potential for toxic residuals. Gas works had coal tar residuals, lumber mills have pentochlorophenol from treated posts and timbers, railway yards may have had chemical leakage into the roadbed, and chemical depots may have stored carcinogenic compounds. The coal tar site became the subject of an argument between the city, the developer, and the upper-tier government as to who would pay for the cleanup. The impact that site reuse planning, or the absence of planning, has on new owners, on insurance costs, and on adjacent property owners can be substantial—that is, in millions of dollars.

As cities in the last century were ringed by industries and the idea of protecting the environment was only dimly perceived, it can be expected that as redevelopment now unfolds, some 50 to 100 years later, redevelopment of many of these sites can be dangerous to workers and to new occupants. However, the proliferation of hazardous chemical compounds after World War II in a poorly regulated environment suggests that as these industries, approaching the 50-year mark at the end of this century, are decommissioned, the next generation of site reuse concerns could have more devastating effects unless extreme caution is exercised. The Love Canal disaster in western New York serves as a prototype of this next generation of problems.

Environmental Protection Phase

The second phase of the environmental management process, environmental protection, is illustrated in detail in the right-hand part of Fig. 3-5. It entails seven modes as discussed, in the Ecoplanning Phase section earlier. Of the six modes, parts of two require some elaboration; the remainder are self-explanatory.

The first item that requires elaboration, the urban and regional government operation mode, is performed at the various levels of governmental jurisdiction. In most countries there are four levels: national or federal, state or provincial, county or metropolitan, and city or municipal.

Secondly, it is necessary to elaborate that, in both the abatement and enforcement stage and the monitoring stage, the evaluation of standards is an upper-tier prerogative, but enforcement of standards may be done at any level, as may rehabilitation and inspection. Matched or shared funding is common; overlapping jurisdictions occur frequently, requiring some dialogue as to which tier becomes the lead agency.

For the environmental protection phase, the development and use of environmental models are helpful in setting ecoplanning priorities. Common models in use are hydrological models detailing dynamics of waves and runoff, lakeshore capacity models, phosphorus models for Great Lakes effluent loading, air quality mixing and loading models, and landscape feature and feature valuation models (Coleman 1975; Fabos and Caswell 1977).

The phosphorus model for effluent control in the Great Lakes is being implemented, with hundreds of millions of dollars already spent on new sewage plants. Argicultural runoff as part of the phosphorus loading issue is the next concern for study and control.

The acid rain models are now being debated both in the scientific community and in the political arenas of the United States and Canada. This model has the potential for generating large economic penalties for certain industries. Models may encourage public involvement and attract public funding. The utility of such models hinges not solely on their cost but on the possibility for verification and for providing correctness, sooner or later, in the real world. If the CO_2 global climate model and the nuclear winter model are similarly successful in changing our actions, we may all have a more secure future—or indeed a future. The model builders are in a sense supertechnocrats, but, leaving this issue aside, their work has been successful in reordering socioeconomic priorities galvanized around specific resource management issues.

The second mode, the new-facility development or old-facility decommissioning mode, has similar characteristics. First, the contract specification writing stage, as far as the environmental aspect is concerned, requires an understanding by lawyers and engineers as to the specific environmental features and processes to be given protection. Contract specifications identify particular mitigation and rehabilitation procedures to be carried out by the contractors. Since these specified procedures are sent out for tender, the bid document should be prepared from the information provided in the preceding EIS and approved by the hearing board. Unless this is done, the EIS does not achieve its purpose but becomes a showcase document used for window dressing at a hearing only.

Futhermore, the contractor prepares a bid on the basis of information provided. Work subsequently required to be done beyond that specified is charged extra at cost plus expenses. In fact, as these extras are often the only real profit made on a job, bids are often made by contractors in the hope that some extras will develop. If the EIS report is appended to the bid document, the contractor is given specific biophysical and cultural data. These data allow for realistic bidding and cost control. However, should there be misinformation in the EIS report, the contractor may have a legal claim for damages against the environmental consultant. An example with which I am familiar centered around a geology map (prepared as part of an EIS) showing no bedrock within the working depth of a pipeline trench. When bedrock was found and required blasting, the blasting costs were claimed by the contractor to the project proponent who in turn sued the environmental consultant for seven-figure damages. In Chapter 6 this example is developed in greater detail.

Construction supervision requires inspectors trained in environmental matters to observe construction activities on a daily or weekly basis to prevent damages to the environment. Saving trees in or near a right-of-way and in or near an excavation site, controlling erosion near streams, reconnecting tile drains to prevent wet spots from developing in farmers' fields, and controlling and stockpiling of toxic chemicals on a construction site are common concerns.

To be effective, the environmental inspector must be familiar with safe operating procedures for equipment, equipment operation specifications (turning radius, angle of stability, depth of engines in stream crossing), environmental permits (burning, cutting, water taking, etc.), and the costs associated with mitigation. In addition, the ability to work long hours, to work in hazardous operating conditions, and to get along with the site superintendent is critical to success.*

The "as-built report," part of the "report stage," documents what happened as the facility was built. Without this documentation, monitoring the effectiveness of the EIS is meaningless. Many alterations to the EIS can take place during construction because of labor strikes, heavy rain, or unfortunate biophysical features such as artesian water, bedrock, or permafrost. The construction adjustments needed to overcome these unforeseen constraints will not be found in the original EIS. From my experience on 38 right-of-way projects, about two-thirds of the EIS mitigation procedures can be anticipated, and may or may not be done, while one-third fall into the unforeseen category. Hence, reliance on the EIS stage to control environmental degradation is simplistic or naive. The EIS literature (see Duinker and Beanlands 1983) generally sidesteps the contract specification writing, construction supervision, and as-built report stages. In teaching impact assessment, university instructors who are deficient on actual construciton experience must rely heavily on theory.

The monitoring stage is a specific action designed to answer specific technical questions. An example is the recovery of benthic organisms in a stream bed following sedimentation resulting from construction. Some authors lump all six stages of environmental protection into the term *monitoring*. This is done out of ignorance of how construction is actually staged and controlled.

Government Policy Formulation Mode

This mode occurs when environmental policies are under review (see left side of Fig. 3-5). Examples of such policy reviews are wetland policy development, food land preservation guidelines, acid rain control policy, and energy policy planning. For aquatic areas, the ecosystem management approach put forward by the International Joint Commission for the

---

*For more details on environmental inspection the reader is referred to two papers, Mutrie (1980) and Mutrie and Dorney (1981).

Great Lakes is significant, as it centers on institutional analysis as well as on a stress-response model for the Great Lakes ecosystems (Great Lakes Science Advisory Board 1982; Francis et al. 1979, 1985). Current resource policy issues in the Canadian context are discussed by Dewiwedi (1980), who looks at issues ranging from northern development to transboundary United States and Canadian resources. To encourage public discussion and debate, briefs and privately prepared position statements can identify government interactions of a policy nature. As well, new ideas are more easily incorporated as more actors participate.

Another function of this mode is preparation of a state-of-the-environment report (SOER). This report, to be prepared at frequent and regular intervals, documents what issues have been and are being addressed and what issues are likely to emerge in the next reporting period. The present lack of such SOERs for lower-tier governments makes it necessary to rediscover continually old issues—to reinvent the wheel. Present socioeconomic statistics published by upper-tier government are generally of minimal utility in measuring trends in environmental quality, as they were not designed with this in mind. An SOER may not only be professionally focused but also can be used to encourage the news media and educators to develop an understanding of social, geographic, and biological issues.

Urban and Regional Government Mode

This mode describes the activities of an environmental planning group for a unit of local government at either the urban or regional scale. As most departments or ministries of the environment constitute an upper tier of government and have jurisdiction over a large region, power flows downward to middle or lower tiers, generally to counties and municipalities. One state attempting multiple-tier integration is New South Wales (Australia) under its Deprtment of Environment and Planning. This department, formed in 1979, includes a regional planning team, a policy unit, and an environmental protection group. It combines the usual separate acts of planning, environmental protection, and environmental assessment into one consolidated administrative unit.

Relatively few lower-tier units of government have an environmental management group. In a case study published by the American Institute of Planners (Linville and Davis 1976), the United States experience at the municipal level was reviewed. The study suggested that a group composed of representatives from various line departments but reporting to the top executive worked best, as it did, for example, in Nashville, Tennessee. In a second situation in San Diego, California, the development of a totally new urban line department of the environment was destroyed by bureaucratic warfare. The Nashville success was apparently the result of middle-level civil servants viewing the environmental secretariat as a prestigious appointment, since the group reported to the top executive level. For this reason the staff dedicated themselves to getting the job done as opposed to viewing the activity as a committee chore of low priority.

At whatever level an environmental unit is organized, such a unit develops and maintains an information storage and retrieval system as part of the environmental analysis. This system is useful to other governmental line departments as well as to the private sector and to the public. Current information on 17 parameters or more would be available (Fig. 3-7) and studies commissioned as needed. The 17 parameters now used in Ontario municipalities are based on a questionnaire developed by Horne (1987). This list is common to many lower-tier governments throughout North America and Europe.

This mode requires further development and evolution by multiple tiers of government. As it is interdisciplinary and draws data from many sources, the mode does not fit easily into a line department structure. Implementation requires a secretariat or service unit available to all units in a government. This unit would be guided by a committee representative of the line units and reporting to the chief administrative officer.

An alternative way to implement this mode is to incorporate or combine it with an expanded planning department, as was done in the New South Wales instance. The nature of planning—its interdisciplinarity and its future orientation—provides a reasonable conceptual an operational fit. By way of conjecture, the problem with this alternative is its vulnerability to pressure which can be brought to bear on the planning group when uncomfortable environmental quality data emerge that have unfortunate or embarrassing socioeconomic and political implications. The New South Wales experience will be interesting to watch.

Corporate Facility Operation Mode

This mode is different from the other five. Its focus is on the day-to-day operation of a facility, including frequent internal redesigns of industrial processes. The technical aspects of environmental planning are handled by a chemical engineer, systems design engineer, or mechanical engineer. Pollution abatement, toxic spills, solid and liquid waste management disposal, and recycling require process analysis, risk analysis, and emergency planning. Many railways and petrochemical companies (see Syratt in Roberts and Roberts 1984, p. 343) have an environmental office or group to handle these matters. If spills occur—for example, when a caustic soda car derails—proper notification and safety procedures must be followed to limit legal liability, and any subsequent prosecution for pollution infractions.

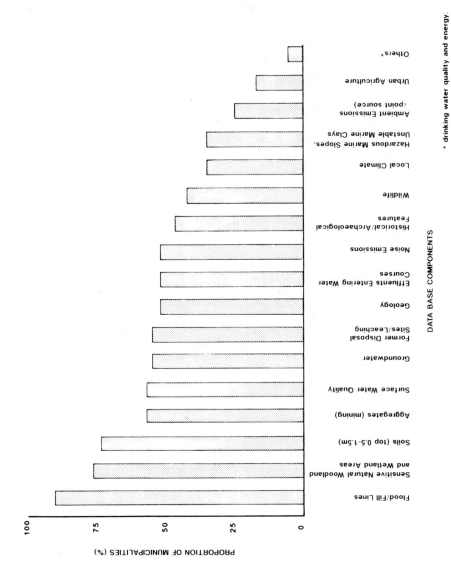

Figure 3-7. Data bases utilized by Ontario municipalities for planning purposes (Horne 1987).

From time to time, specialists in organic chemistry, limnology, and toxicology are contracted by industries to work on specific waste disposal issues. Examples are a situation where greases in a valve station are emitting toxic fumes affecting livestock or where a change in the government's allowable effluent standards for a paper mill requiring some bioassay work in fish under controlled laboratory conditions. Also a common example is a pollution study in the factory or mine where health risks are suspected, particularly where health and safety standards may be exceeded.

Another, somewhat different aspect of corporate facility operation mode concerns the mining industry. Tailings, overburden, and mine wastes require special management. For open-pit mining, such as gravel pits or coal mining, progressive rehabilitation plans are now being prepared (McLellan 1984). Occasionally, attempts at rehabilitation are of questionable value from both economic and biological points of view (Dorney 1984).

The environmental manager with a biological or geographic background lacks the special skills in organic chemistry and toxicology. Hence he or she is less involved in the corporate facility operation mode than in the other modes. Not infrequently, however, an environmental manager may be called on to recommend specialists or to provide advice from a generalist's perspective to complement the advice offered by the specialists. Because of prosecution for violations of pollution laws, confidentiality between a client and a consultant poses particular problems in this mode.

Comparison of Ecoplanning and Protection Approaches

Ecoplanning and protection, as described here, evolved to cover situations likely to be encountered in the Great Lakes region of the United States and Canada. These processes have the greatest similarities to the environmental planning procedures of Kragh (1967) in Germany. Kragh outlines the following steps: survey, analysis, evaluation, policy, design, and implementation. The phases, modes, and stages depicted in Fig. 3-5 embrace these general categories and specifically itemize administration, environmental impact, contract specification writing, environmental inspection, various reports, aesthetics, renovation, public participation, and communications, thus fleshing out Kragh's concept.

Hackett (1971) describes a concept of environmental planning that is based on new facility development. His process proceeds from survey to analysis to proposals, somewhat similar to the four stages of ecoplanning in Fig. 3-5 for the new facility development mode. However, my concept of ecoplanning goes further by specifying five other modes in which environmental advice is offered, and it includes an environmental protection phase. In part this reflects the growth in the practice of environmental management since his book was written.

The environmental planning approach advocated by Kade (1972) has three elements: overall disturbance in the ecosystem, organization of the planning process, and goal fixing. Little explicit concern for systematic resource mapping is indicated. The process whereby data are synthesized and systems properties are defined is bypassed. Placing goals last appears to be the reverse of most planning processes in which Kade's approach seems rather sketchy, although his publication may not do justice to his ideas.

McHarg (1966) outlines six steps for environmental planning:

1. Ecological inventory (geology, climate, water resources, soils, biota, etc.)
2. Historical inventory
3. Description of natural process
4. Description of limiting factors (hazard lands, lack of water, growing season, etc.)
5. Attribution of relative values
6. Description of intrinsic permissiveness and prohibitions to prospective land use (land capability and suitability).

This approach is suitable for the urban and regional development mode, the new-facility mode, and perhaps the old-facility decommissioning mode, but not the others. Also, the conceptual framework appears to sidestep the entire environmental protection phase upon which so much successful planning depends.

In a later paper McHarg (1967) utilized the landscape model of contrasting "health" or "ill-health" as a way of conceptualizing the planning approach to be followed in urban and regional areas. Under "health," McHarg includes complexity, diversity, interdependence, stability, high number of species and low entropy; "ill-health" is the opposite of these terms. Although this binary approach appears to be logical and supportable on conceptual grounds, how it might be implemented is not clear. The analogy of health in man and animals and health in landscape needs an explicit methodological framework. For example, if all cultural landscapes were low-entropy, diverse, stable, complex ecosystems, our urban and industrial lifestyles, which depend on high production from monocultural crop systems, would not be sustainable. In addition, such biologically simplistic agriculture and urban systems need uniformity and redundancy to capture the social energy efficiently (investment, commerce, communications, decision making).

Hills, et al. (1979) provide a conceptual basis for ecological land use planning based on five perspectives:

1. The nature of making decisions
2. Transactional behavior in a systems context

3. Use of gradient land classes having physiographic, biotic, and cultural features for purposes of landscape classification and evaluation
4. Differentiation of ecosystems on the basis of homogeneity
5. Development of human ecosystems that are controlled by an integration of public welfare, institutional, and natural environmental features.

No ecologist can quarrel with these perspectives. They are not, in my judgment, sufficiently related to historical process, cultural differences, new technology, and the variety of existing land uses and potential land management strategies. Different teams of specialists using this approach may come to widely divergent conclusions, suggesting that the methodology is only helpful in scenario development, not in optimization, as the reader is led to believe.

The underlying assumption in applying the Hills' approach is that land capability, or its ability to produce organic matter (point 3), should become the major determinant in land use allocation. This assumption, if followed religiously, may be too simplistic. Other authors, as Miklos et al. (1986) in Czechoslovakia, use an optimization technique based on scientific principles. In my opinion, land use planning, if it is to be effective in a politically open arena, must take account of human perceptions and historical uses of land. Viewed from this time perspective using a landscape evolution mode (Fig. 3-2), competing present and future uses for land can be judged on their respective merits. Futhermore, new technologies, fluctuating economic circumstances, and varying human perceptions must be understood, argued on merit, and accommodated in the final analysis. In my opinion, there may never be one single best land use plan or one single best land use planning process. In fact, the master planning approach to environmental land use planning, which tends to result from the approach suggested by Hills, et al. (1979) and Miklos et al. (1986), is rigid and can become too technocratic to be implemented in an open society. If such planning were mandated, it could stifle innovation, reduce incentive, and severely stress the democratic principles upon which representative government rests. The six environmental planning modes (Fig. 3-5), although comprehensive, deliberately attempt to be more flexible to downplay the role of the technocrat and to underscore the need for considerable public diologue.

Only Crowley has developed a method of spatially classifying land in terms of a hierarchy having three levels (Cassie et al. 1970). The smallest unit is the *stand* or *geotype* (a soil parcel and its characteristic climax plant cover); the intermediate level is the *site* (a soil type with any number of potential plant communities); and the third level is the *catena*. This system is not meant to form the conceptual framework for environmental planning, but it is potentially a useful technical tool, having strong similarities to the Hills land capability approach and to what is called ecological land classification. One drawback in its use is that it is hard to explain to the public, news media, and politicians.

Dansereau (1972) has proposed an ecological land occupation (classification) system that relates various trophic levels to two social levels, *investment* and *control*. Translated into colored mapping, historical processes can be deduced as land use changes having occurred over time. The application of the Dansereau system attempts to link energy processes between natural and human systems to specific land use parcels. It may prove in time to be a useful environmental planning tool. However, one of its drawbacks is the complexity in the mapping system, the cost in doing the mapping and the updates, and the difficulty the uninitiated have in learning the system. Since aerial photo data are required to do the mapping, time frames before World War II are not available. It was for all these reasons that the landscape evolution approach was developed.

Moss (1984) suggests an ecological process orientation relying on energy and moisture as key elements driving natural and cultural landscape systems. With this approach, maps depicting primary productivity, for example, at a scale of 1:1,000,000, can be developed. Such graphic representations are useful for making some kinds of regional development decisions. The concept of developing ecological process maps requires further experimentation to see if it can be used for more than just the regional development decisions Moss has illustrated.

Glikson's (1967) contribution to environmental planning is generally overlooked. He lists various landscape functions to be integrated into regional landscape planning. They include sustained intensive production, habitability and attractivity, recreation, and manifestation of geological and cultural continuity. It is not a matter of society or nature, but the necessity to establish "social and biological controls to ensure the survival, usefulness and value of landscape as a life resource." Presumably this distinguishes land as a commodity to be consumed from land a resource to be husbanded. As an early statement of what landscape planning should be, his work is insightful.

The ABC approach (Table 3-1) has the advantage of simplicity, replicability, and variety of scales. The simplicity enhances its ease in communication to any technical or lay audience. Its replicability decreases the chances for landscape interpretation being challenged before political bodies and hearing bodies and shifts the focus from the analytical stage to where it belongs—at the evaluation and design stage. The variety of scales for depicting information gives this methodology flexibility. The ABC methodology, however, is a static-feature mapping methodology requiring supplementary graphics and tabular material to elucidate historical and future ecological processes. It can be used as an overlay (sieve mapping) system, but its simplicity (only three maps) has meant this is seldom necessary.

The ABC methodlogy has been tested for regional planning purposes in the Yukon by J. Bastedo (1982) and B. Hans Bastedo (1983). They suggest that it can be effective in such extensive natural ecosystems as well as in the agroecosystems and urban ecosystems where it was first used (Dorney 1977).

Any of the mapping systems described can be digitized, stored, and displayed by computer-aided plotting systems. The cost effectiveness of using computer mapping systems depends on the type of questions to be answered—that is, on the particular application and on the particular technology being used. Besides the cost of providing digitized information for graphics display, my experience has been poor in using such graphic systems in a small planning practice. The problem is one of cost overruns, difficulties in displaying easily understood graphics, meeting targeted deadlines, and managing personnel. The last problem ranges from apparent sabotage of tapes when an employee's contract has terminated, unauthorized erasing of tapes, and costly continual debugging resulting from tampering or program modification. A permanent computer graphics staff is mandatory with access controlled carefully by a few senior professionals. Upper tiers of government and occasional users find computer-driven landscape systems worthwhile as the technology is becoming simpler and less expensive.

This survey of current approachs to environmental planning shows how the field of environmental planning, protection, and management has evolved over the past decade and a half.* Most authors see environmental planning as both a technical and a political process in which society and nature are brought together. Furthermore, all see the need to adopt a systems approach capable of identifying landscape processes, such as feedbacks, thresholds, lags, stability, resiliency, linear and nonlinear properties, and various dynamic socioeconomic and cultural relationships. By the very nature of adopting the systems approach, environmental management requires an interdisciplinary approach.

Conceptual Anomalies and Research Needs

Given the rapid 15- to 20-year evolution of environmental management, it is not surprising if definitions of terms, the modus operandi, and communication among academics and practitioners are less than perfect. If we look ahead, however, some new dimensions are emerging.

The use of environmental assessment by the courts and by citizen groups to stop projects, although effective at times, provides little leverage in overall landscape evolution. The major levers to control unwanted change lie in doing more analysis on emerging technologies, on the large urban and regional development initiatives, and on policy

---

*For additional references, see the *Environmental Planning Resource Book* (Lange and Armour 1980).

initiatives that set the stage for subsequent specific facilities. To rely principally on environmental assessment to maintain environmental quality relegates environmental management to an ad hoc approach.

The need for more experience and documentation on all aspects of the environmental protection phase seems evident. Unless planners develop a sensitivity to the implementation of their work, planning cannot improve in a step-by-step manner based on actual experience. The use of as-built environmental reports and environmental audits offers promise for employment and effective corporate control. Such documents enable courts to asses penalties and damages, to make corporates endeavors more socially and environmentally sensitive, and to control and monitor internal actions.

New satellite multispectral scanners offer promise for updating land use inexpensively. If such digitized electronic sensors can be coupled to various land process models, such as to a land use erosion model, environmental protection should be easier to accomplish. Current trends in productivity can be tracked without waiting years for census data to be published. The contributions of Coleman (1975), Fabos and Caswell (1977), and Miklos et al. (1986) in developing computerized landscape analysis systems provide useful input, if placed in a human ecology context with open and continual public participation.

The issues of public health are always paramount. As we become more aware of the linkages between human health and waste, air, food, and water supplies, enormous pressures are being generated for effective environmental management. Issues such as dioxin in drinking water or leachate plumes from landfill sites adjacent to housing areas have the potential to disrupt land uses with attendant socioeconomic dislocations. If professional flexibility is maintained, human resources with a high managerial ability can meet the seen and unforeseen challenges. By developing at the municipal-level SOER, which includes health parameters, and by integrating health with other social and biophysical concerns, delivery of better environmental quality should be facilitated. Broadening the environmental mandate indicates we are moving toward a much greater respect for intrinsic ecological values.

Bibliography

Balser D, Bielak A, deBoer G, Tobias T, Adindu G, Dorney RS (1981) Nature reserve designation in a cultural landscape, incorporating island biogeography theory. Lands Plan 8:329–347.

Bastedo J (1982) A resource survey method for environmentally significant areas (ESA) in the Yukon with results from the Bennett Lake/Carcross Dunes/Tagish Lake ESA. Unpublished master's thesis, School of Urban and Regional Planning, University of Waterloo (Ontario).

Beccasio AD et al. (1980) Atlantic Coast Ecological Inventory User's Guide and

Information Base. Washington: Biology Services Program, U.S. Fish and Wildlife Service.

Brady RF, Tobias T, Eagles PFJ, Ohrner R, Micak J, Veale B, Dorney RS (1979) A typology for the urban ecosystem and its relationship to larger biogeographical landscape units. Urban Ecol 4:11–48.

Canada Land Inventory (1969–1977), Department of Regional Economic Expansion, Ottawa.

Cassie DR, Coleman DJ, Howard JH, Veillette J, Crowley JM (1970) Geography of Ecosystems in South Wellington County, Ontario. University of Waterloo (Ontario): Division of Environmental Studies.

Chanasyk V (1970) The Haldimand-Norfolk Environmental Appraisal. Toronto: Ontario Ministry of Treasury, Economics and Intergovernmental Affairs, Vols 1, 2.

Coleman DJ, MacNaughton I (1971) Environmental planning in Waterloo County. In: McLellan AG (ed) The Waterloo County Area Selected Geographical Essays. Waterloo, Ont.: University of Waterloo, Geography Department.

Coleman DJ (1975) An Ecological Input to Regional Planning. Waterloo, Ont.: University of Waterloo, School of Urban and Regional Planning.

Dansereau P (1972) Biogeographie dynamique du Quebec. In: Grenier F (ed) Etudes dur la Geographie du Canada. Tornoto: University Press.

Dansereau P, Pare G (1977) Ecological Grading and Classification of Land-Occupation and Land-Use Mosaics. Ottawa: Geographic paper No. 58, Fisheries and Environment Canada.

Dewiwedi OP (ed) (1980) Resources and the Environment: Policy Perspective for Canada. Toronto: McClelland and Stewart.

Dolan R, Lins H (1987) Beaches and barrier islands. Sci Am 257 (1):68–77.

Dorney RS (1977) Biophysical and cultural-historic land classification and mapping for Canadian urban and urbanizing land. In: Thie J, Ironside G (eds) Proc Workshop on Ecological Land Classification, Series 3. Ottawa: Environment Canada, pp 57–71.

Dorney RS, (1984) Rehabilitation: is the cure worse than the disease. Lands Arch (May) p 120

Dorney RS (1985) Predicting environmental impacts of land-use development projects. In; Whitney JBR, MacLaren VW (eds) Environmental Impact Assessment: The Canadian Experience, Environmental monograph No. 5, publication No. EM-5. Toronto: University Press, pp 135–149.

Dorney RS, Evered B, Kitchen CM (1986) The effects of tree conservation in the urbanizing fringe of southern Ontario. Urban Ecol 9:289–308.

Dorney RS, Giesbrecht H, et al. (1969) Ecoplanning—The Contribution of Ecology to the Process of Urban Planning. Waterloo, Ont.: University of Waterloo, School of Urban and Regional Planning.

Dorney RS, Rich SG (1976) Urban design in the context of achieving environmental quality through ecosystems analysis. Contact 8:28–48.

Dorney RS, Smith L (eds) (1985) Environmental Mediation. Working paper No. 19. Waterloo, Ont.: University of Waterloo, School of Urban and Regional Planning.

Dorney RS, Wagner-McLellan PW (1984) The urban ecosystem: its spatial structure, its scale relationships, and its system attritures. Environments 16(1):9–20.

Duinker PN, Beanlands GE (1983) Ecology and Environmental Impact Assessment: An Annotated Bibliography. Halifax: Institute for Resource and Environmental Studies, Dalhousie University.

Fabos JG, Caswell SJ (1977) Composite Landscape Assessment. Research Bulletin 637. Amherst: University of Massachusetts Press.

Fisheries and Oceans (Canada) and Ministry Natural Resources (Ontario) (1981) Great Lakes Shore Management Guide. Ottawa: Environment Canada.

Francis GR, Magnuson JJ, Regier HA, Talhelm DR (1979) Rehabilitating Great Lakes Ecosystems. Technical Report 37. Ann Arbor, MI: Great Lakes Fishery Commission.

Francis GR, Grima APL, Regier HA, Whillans TH (1985) A Prospectus for the Management of the Long Point Ecosystem. Technical Report 43. Ann Arbor, MI: Great Lakes Fishery Commission.

Garlauskas AB (1975) Conceptual framework of environmental management. J Environ Mgmt 3:185–203.

Gierman DM (1981) Land Use Classification for Land Use Monitoring. Working paper No. 17. Ottawa: Lands Directorate, Environment Canada.

Glikson A (1967) The relationship between landscape planning and regional planning. In; Towards a New Relationship of Man and Nature in Temperate Lands, Part II: Town and Country Planning Problems. ICUN Publ New Series No. 8, Morges pp 37–50.

Great Lakes Science Advisory Board (1982) Annual Report. Windsor, Ont. Great Lakes Research Review.

Hackett B (1971) Environmental Planning. Newcastle upon Tyne, England: Oriel Press.

Hans Bastedo B (1983) Cultural aspects of resource surveys, a human ecological approach: Aishikid, Yukon. Unpublished master's thesis, University of Waterloo (Ontario) School of Urban and Regional Planning.

Hills GA (1974) A philosophical approach to landscape planning. Lands Plan 1:339–371.

Hills GA, Love DV, Lacate DS (1970) Developing a Better Environment. Toronto: Ontario Economic Council.

Horne R (1987) Environmenal co-ordinators in the Ontario municipal government structure. Unpublished master's thesis, University of Waterloo (Ontario), School of Urban and Regional Planning.

Kade G (1970) Introduction: Economics of pollution and interdisciplinary approach to environmental planning. Soc Sci J 22;563–575.

Kragh G (1967) The components of landscape planning: Survey, analysis, evaluation, policy, design solution, implementation. In: Towards a New Relationship of Man and Nature in Temperate Lands, Part II: Town and Country Planning Problems. IUCN Publ new series No. 8, Morges, pp 90–95.

Lang R, Armour A (1980) Environmental Planning Resource Book. Ottawa: Land Directorate, Environment Canada.

Linville J Jr, Davis R (1976) The Political Environment: An Ecosystems Approach to Urban Management. Washington: American Institute of Planners.

McHarg I (1966) The future of environmental improvement: Reaction. In: Environmental Improvement (Air, Water and Soil). Washington: U.S. Department of Agriculture Graduate School.

McHarg I (1967) An ecological method. Lands Arch 57(2)105–107.

McHarg I (1969) Design with Nature. Garden City, NY: Natural History Press.

McLean B (1986) Site Re-Use: Concerns of Developing Polluted Land. Senior Honours Essay. Waterloo, Ont.: University of Waterloo, School of Urban and Regional Planning.

McLellan AG (1984) Monitoring and modelling progressive rehabilitation in aggregate mining. Bull Int Assoc Eng Geol 29:279–283.

Miklos L, Miklisova D, Rehakova Z (1986) Systemization and automatization of decision-making process in LANDEP method. Ekologia (CSSR)5(2):203–232.

Moss MR (1983) Landscape synthesis, landscape processes and land classification, some theoretical and methodological issues. Geojournal 7(2):145–153.

Moss MR (1984) Environmental Process Data Inputs to Systems of Land Classification: Ecoregions and Ecodistricts of Ontario and Manitoba. Guelph, Ont.: University of Guelph, Department of Geography.

Mutrie DF (1980) The Leopold inventory technique—a critical evaluation and case study of the east side of Lake Winnipeg. Unpublished master's thesis, University of Waterloo (Ontario), School of Urban and Regional Planning.

Mutrie DF, Dorney RS (1981) Environmental concerns in rights-of-way management. Proceedings of Second Symposium on Electric Power Research Institute, Palo Alto, CA (Box 50490), pp36-1–36-9.

Nemerow NL (1985) Stream, Lake, Estuary, and Ocean Pollution. New York: Van Nostrand Reinhold.

New South Wales (1979) A Guide to the Environmental Planning Legislation. Canberra, Australia: Department of Environment and Planning.

Pikey OH Sr, Pikey WD, Pikey OH Jr, Neal WJ (1983) Coastal Design. New York: Van Nostrand Reinhold.

Potomac Planning Task Force (1967) The Potomac: A Report on Its Imperiled Future and a Guide to Its Orderly Development. Washington: Department of the Interior.

Roberts RD, Roberts TM (1984) Planning and Ecology. London: Chapman and Hall.

van Classen RDB, Ross GA (1981) Landsat for Resource Evaluation and Management in the Alberta Foothills. Winnipeg: 7th Canadian Symposium on Remote Sensing.

4

Organization and Development of a Private Practice in Environmental Management

Developing and organizing a consulting practice in environmental management has features distinctive from other professional ventures. These features center on the need for undertaking and integrating interdisciplinary studies* and the resulting complexity of communication between professionals, scientists and technicians, and the public. The management of interdisciplinary activity requires explicit, close attention without which little can be achieved that cannot already be done in conventionally or traditionally organized consulting efforts based on a single discipline. Further, the needed for clear and accurate communication is more acute than in single-discipline endeavors.

First, there is the issue of minimum or critical size to carry out interdisciplinary work. I believe an environmental management practice, whatever its focus, cannot be organized and operated easily by one person, unless technically competent specialists are readily available as subconsultants to form a team.† The complexities of the human and natural ecosystem, be it terrestrial or aquatic, temperate or tropical,

---

*Interdisciplinary, as defined by M. Guy Berger (Apostel et al. 1972, pp 26–27), describes the interaction between two or more different disciplines which may range from simple communication of ideas to mutual integration of organizing concepts, methodology, procedures, epistemology, terminology, data, organization of research, and education in a fairly large field.

†An interdisciplinary group or team "consists of persons trained in different fields of knowledge with different concepts, methods and data and terms organized into a common problem with continuous intercommunication among participants from different disciplines" (Apostel et al. 1972, pp 26–27).

human-dominated or devoid of such impacts, and the voluminous technical literature in limnology, fisheries, wildlife management, soils, geology,
plant ecology, waste management, restoration ecology, and the social
sciences make it difficult for one peron to absorb without the considerable
technical depth provided by a more specialized team or specialized
subconsultants. For ease in day-to-day operations, as backup in case of
illness or holidays, and for reasons of cost effectiveness, some members
of an interdisciplinary team should be available in house. A firm of two to
four senior professionals whose natural, geographic, and social science
discipline backgrounds reinforce each other is a core group to work in
environmental assessment and the facility development mode. For such a
core I would select a forester or plant ecologist, a vertebrate animal
ecologist or wildlife management scientist, a soil scientist or physical
geographer, and a limnologist or fisheries scientist.

   To this core could be added a design or engineering component and a
social science component. To work in environmental protection, principally waste management, a hydrogeological engineer, toxicologist (perhaps with a chemistry degree), transportation engineer, planner, and an
ecologist–landscape management specialist might comprise the core. A
coordinator to organize and synthesize the technical disciplines and to
operate the projects cements the mix together.

   The use of freelance subconsultants to round out the team effort
effectively reduces overhead costs. Not uncommonly, during the preparation of a proposal, freelance consultants may be committed to other
groups. To locate equivalent replacements is a tedious, time-consuming
process. Keeping an up-to-date list of freelance specialists is useful. This
list may include university staff or retired professionals who wish to work
on an as-needed basis, adding credibility and making a personal network
available to young professionals.

   In addition to size there is the interdisciplinary issue. There are four
approaches to this issue. The first is to organize a firm whose principals
are all natural science trained. The second is to combine scientists with
designers—that is, with planners, landscape architects, and architects.
The third is to combine personnel having engineering training with
personnel having environmental science backgrounds. The fourth is to
mix social scientists with these other combinations. Since each approach
is rather different, it seems worthwhile to explore the possible advantages
and disadvantages of these four interdisciplinary approaches without
suggesting that one approach is better than another.

Various Discipline Backgrounds of Principals in the Firm

A Science-Only Staff

The advantage of this arrangement—irrespective of whether the principals' backgrounds are in plant ecology, forestry, animal ecology,

wildlife management and soils, geomorphology or geology—is that all personnel have had some training in common, for example, in mathematics, biology, chemistry, physics, earth sciences. Thus they have a common philosophical and problem-solving orientation. This commonality facilitates group projects, allowing personnel to integrate their work from a conceptual and methodological point of view rapidly and efficiently.

The disadvantages with this approach are two. First, the specialized scientific language or jargon is a barrier to interprofessional and public understanding. Furthermore, few scientists have any training in graphic methods or public speaking. As this inability hampers professional communication, credibility to public groups, elected politicians, and hearing boards may be slow to emerge. The inability to communicate clearly and forcefully by an otherwise competent scientist results in lawyers' pressing their clients to select other professionals as effective witnesses. Weak communication skills ultimately limit a group's chances for financial success as a firm. As scientists compete for the dollar in a crowded professional arena, more attention has to be paid to the ability to communicate by on-the-job training workshops stressing public speaking and graphic presentations.

The second disadvantage is that the rational scientific method is only one way to develop solutions to complex planning problems in ecosystem management. Intuitive solutions, design solutions, public dialogue to develop solutions that are politically acceptable, just to mention a few, often irritate those who assume that rationality should dominate every problem-solving situation. An effective team effort may combine planning processes, resulting in a process called "mixed scanning."

A Science-Design Staff

The advantages in this staff combination are many. First, subcontract work with other planning, landscape architecture, or architecture firms is easier to undertake, since at least one member of the firm has some common university experience, common professional society membership, and common language with counterparts in other design firms. A lone designer among a group of scientists is not a competitive threat to a design-based firm.

Second, by understanding the planning process, which includes administrative, legal, and policy aspects, design concerns can be dovetailed more sensitively with scientific analysis. It is often in these administrative, legal, and policy matters that projects run into serious difficulties and frustrate scientists who may poorly perceive the reality of the sociopolitical environment. For example, a change in policy by a government agency, planning board, or elected council midway through an environmental impact study may necessitate considerable change in the field analysis and possibly a renegotiation of the budget. Under these circumstances it is helpful to have in a firm someone attuned to the situation. Understandably, these are skills that can be learned by anyone in the school of hard knocks, but since credibility is vital to professional survival, too many hard knocks are disastrous.

A third advantage is the in-house experience in professional planning that comes with having worked with various community groups, politicians, and other professionals. This is advantageous to any professional office dealing with construction impact, construction supervision, and construction-design problems. Urban, urbanizing, and rural areas have both biological functions that can be measured and economic, social, legal, and political realities. Well-organized public agencies, private corporations, and their related professionals deal in these realities and can either be helpful, neutral, or hostile to an environmental planning firm. Although it is time-consuming to link planning and the sciences, an individual with a graduate degree in science combined with graduate or advanced training in a professional planning school can bridge these two professional areas.

One disadvantage of including someone with a planning-design degree in a firm oriented toward ecological or scientific analysis is the difficulty in finding such a hybrid individual who has a working knowledge of both areas and a desire or motivation to orient a career in this direction. As planning and landscape architecture schools in Canada and the United States continue mixing ecological science and planning-design training, such hybrids will enter the labor market. The converse, that of science-oriented environmental studies programs providing an awareness of planning or design issues from a social, economic, and political perspective will also develop the hybrid scientist-planner.

A second disadvantage in having a science-based ecological firm hiring a planner is that a person trained as a planner may not find his or her time fully occupied; hence, this share of the workload may be less than that of the science personnel and create staff inefficiencies.

In any case, a firm represented by a mix of professional planning and design and scientific expertise is a workable arrangement. I have found this combiniation to be satisfying to both the planning and design–oriented person and the science-oriented person. Also, this mix is mutually reinforcing, because it potentially has a broader conceptual base—as long as interpersonal communications remain excellent and each discipline respects the other.

A Science-Engineering Staff

The desire of engineering firms to have some ecological science or environmental management expertise in house is a third organizational form now quite common, and there are some advantages to this. When an engineering problem is in its early conceptual stages, it is easier to pop across the hall and ask "Jane" or "Joe" about possible environmental interactions than it is to call in a consultant. Overhead costs (library facilities, secretarial support, drafting, duplicating facilities) are usually shared, as they are in many kinds of professional offices. An engineering firm containing environmental science can advertise that it has a more complete analytical package in house to provide better or more complete service to clients. Cross-fertilization between disciplines is encouraged by having environmental scientists or mangers embedded within a group of engineers; discipline reinforcement takes place by having a hydrologist with a civil engineering background easily accessible to a limnologist, or a soils engineer accessible to a forester–soils scientist.

The disadvantages, on the other hand, of such an ecology/environmental management and engineering combination are subtle but must be recognized before either discipline group decides to combine its efforts. First, environmental scientists, who are generally younger, often find it difficult to criticize senior engineering members of the same firm. Even though environmental scientists dislike the conceptual design of a highway, a sewer network, etc., because of increasing the environmental disruption, they may be less than candid about their concerns. This reticence creates hostility and reduces the openness and spontaneity necessary for trans- and interdisciplinary work of high quality and innovation.

Second, public groups often are suspicious of the objectivity of environmental scientists working under the direction of an engineering firm. To many environmental action groups, the engineers are the "bad guys"; mixing the environmental "good guys" in with them suggests collusion—naive perhaps, but perceived this way. In addition, as engineers derive considerable fees from construction supervision, environmental scientists, who may suggest a no-build alternative or a less costly solution, in fact are reducing income for the firm. In theory, this should not affect sound professional judgment, but readers can judge if they believe it might occasionally apply subtle pressure to a junior scientist.

Third, if the engineering paradigm is reductionist, it clashes with the systems perspective of the environmental manager. In my judgment the environmental paradigm is first and foremost a systems perspective (see Fig. 1-3). A blend of the two approaches leads to better understanding and potentially more creative solutions, but blending is not always smooth.

Last, the ideological conflicts that often arise between an engineer who enjoys seeing something built and an environmental manager who may not can cause tension between the two discipline areas. In an independent firm whenever such conflicts arise, and they are not infrequent, further joint work, if it is done at all, is done under more carefully controlled conditions by the key principals. A controlled condition decreases exterior tension, but in-house tension can cause an environmental manager to leave the firm. This, too, is not an uncommon situation. Some large engineering firms with which I am familiar went through three to five complete science groups between 1972 and 1987.

A Social Science Component in a Mixed-Discipline Firm

In theory this combination is capable of greater creativity and higher efficiency. However, I have never seen it tried. Economists, sociologists, and political science–based firms are already well established, making it easier to collaborate with them than to duplicate their services. Because of conceptual and linguistic differences, coordination and integration with too broadly based a group may become tedious. The challenges are apparent.

In summary, I do not believe there should be a strong preference shown for any one of the four possible approaches. They each have their strengths and weaknesses. However, a single environmental manager considering joining an engineering firm should carefully assess the ideological differences before engaging in such a venture. In my experience, some engineering and some design firms are quite amenable to environmental science input, while others are uncertain or openly hostile.

Legal Arrangements

There are numerous legal arrangements for establishing a firm that need careful consideration, since each arrangement has certain advantages and disadvantages. The issues addressed here are only meant to help or to assist in the dialogue with legal counsel, not to offer legal advice.

First, those contemplating a joint venture with an established engineering firm or planning firm, which usually has a much larger staff, may consider maintaining a separate corporate name and selling no more than 49 percent of their stock to the larger firm. To some extent a separate identity is assured, while a fair amount of work derived from the larger firm is "built in" to provide economic stability. In addition, useful contacts with clients can be more quickly established, a process that takes 2 to 5 years without this initial help. If the parent firm provides a member to the board of directors, this enriches the experience of the board. In addition, the working capital the larger firm has available allows for better budgeting and overall fiscal stability.

The disadvantage in this arrangement is that firms in direct competition with the parent firm are reluctant to use a subsidiary, even if it is less than 50 percent owned. This arrangement, with a possible exchange of stock with a larger engineering firm, is illustrated in Table 4-1.

The second arrangement, setting out the working arrangements in a memorandum of understanding, links firms without an exchange of stock and with or without mutual board membership. This more informal

Table 4-1. Hypothetical organization of a semiindependent environmental management firm tied to an engineering consulting group.

Environmental management firm[a]	
Professional staff (N=9)	Senior environmental manager (1)
	Soil scientist (1)
	Plant ecologist (1) or forester (1)
	Animal ecologist (1)
	Aquatic ecologist (1)
	Support staff, business manager (1)
	Typists (2)
	Communications specialist (part-time)
Board of directors (N=5)	Senior environmental manager, chairman (1)
	Business manager, secretary (1)
	Plant ecologist (1)
	Soil scientist (1)
	Representative of engineering firm (1)
Engineering firm	
Professional staff (N=75)	Environmental engineer (25)
	Hydrologist (25)
	Civil engineer, etc. (25)
Support staff (N=15)	

[a] Shares for the environmental firm may be distributed as follows: Environmental manager, 100; Soil scientist, 100; Plant ecologist, 100; Business manager, 100; Engineering firm, 200; and Treasury shares, 1,400, for a total of 2,000 authorized shares (no par value).

working relationship can improve fiscal stability without heavy legal red tape.

As to formal legal identity for an environmental management practice, various possibilities, such as a partnership, an unorganized group of associates, or a corporation (profit or nonprofit), are worth exploring with legal counsel. Since local tax laws, liability laws, and laws relating to professional corporations and bankruptcy vary widely from place to place, they should be examined fully. Also there are certain pension and estate tax benefits that corporations and partnerships may enjoy. All these aspects should be carefully discussed with legal counsel before a decision is made as to what organizational approach should be used.

We chose to be a private (for-profit) corporation, since it allowed flexibility to sell stock to those whom we might wish to bring into the firm. It also gives us control of sales of stock to heirs or to those leaving the firm by provisions in the articles of incorporation and shareholders' agreements. Laws regulating nonprofit corporations are usually not as flexible for a consultant as laws relating to a for-profit corporation.

Nonprofit firms must dispose of profits within a few years, making it more difficult to build equity. Hence, the nonprofit approach was rejected, as equity improves stability when markets are soft.

Scope of Services Between an Environmental Science–Based Firm and Other Professional Firms

The scope of services a firm offers affects staffing, cash flow, and the kinds of clients it attracts. Unless organized with many senior people who can capitalize the firm, the firm has to be content with many smaller jobs in the role of a subconsultant and gradually develop a recognized competence to take on larger tasks. To be invited into joint corporate ventures or consortiums where all firms are more or less equal in contribution is not easily achieved at the outset. Such invitations arise after some years of experience have shown the ability of the environmental management firm to contribute effectively in preparation of the proposal, to be convincing in presentations to the prospective client, and to work effectively with key members of other disciplines and the firms they represent.

Rather, it is much more frequent to be a subcontractor to another large firm on environmental matters, either reporting jointly to the client or reporting through the prime consultant, who then transmits environmental information to the client. When reporting to the client indirectly through a prime consultant, the environmental consultant must ensure that the environmental information is not edited out or incomplete. The environmental consultant's reputation can be severely damaged if the prime consultant has inadvertently eliminated from his or her report an essential part of the environmental information. As a subcontractor, a small environmental firm usually finds itself asked to undertake the following tasks for a prime contractor (Table 4-2). Although the list in Table 4-2 is only illustrative, it demonstrates the pattern of activities in which environmental firms are engaged.

If the environmental work is for a government agency, it is not unusual to find a section of the agency that has an engineering, an architectural, a landscape architectural, or a planning group equivalent to the consulting counterparts described above. The work may cover the conceptual or feasibility design-planning stage or the master planning-design stage, or it may involve preparing guidelines for construction and monitoring. If it is a hearing or court case, environmental work involves assessing the work of others and testifying as to inadequacies or differences in interpretation of the same environmental data. Such hearings, because of their adversarial nature, are psychologically demanding, but never lacking in excitement.

Table 4-2. Ecological assistance undertaken by small environmental firm for a prime contractor.

Prime contractor	Ecological assistance required
Civil engineering firms	Environmental impact of highways, dams, marinas, industrial parks, landfills
Architecture and landscape architectural firms	Design assistance for small housing sites Design assistance for special use areas (such as zoos, institutional buildings) Recommendations on the retention or renovation of native vegetation for construction sites Identification of unique faunistic and floristic resources Assessment of the quality of aquatic resources Assessment of potential eutrophication of ponds, urban lakes Determination of environmental carrying capacity for recreational uses or parks Completion of tree inventories and arborists' reports Recommendations for site rehabilitation (restoration ecology)
Planning firms	Environmental quality mapping at varying scales (usually between 1:50,000 and 1:5,000) Land capability or land suitability mapping for urban development Open space suitability analysis Impact of alternative urban designs on natural systems (usually housing in parcels from 100 acres or 40.48 hectares to 20,000 acres or 8,091.16 hectares at a scale of 1:5,000)
Law firms	Expert testimony for hearings, court proceedings on issues of environmental impact, damage or potential damage due to pollutants, land value for expropriation purposes and land rehabilitation Meetings with citizens' groups, politicians in informal or formal settings

Discipline Linkages Between Environmental Sciences and Social Sciences

The complexity of any environmental impact study and land capability and suitability study derives as much from the historic human use of land, the value systems of residents living in the area, the capability of the institutions, and the regulatory-legal aspect as it does from the biophysical resources. For this reason, successful environmental planning and protection by a professional environmental manager require that the individual be conversant with a number of social science discipline areas. By *require,* I do not mean that exhaustive social sciences studies are needed, but at least a credible examination and understanding of their relevance are required. The following list, derived in part from Chapter 2, illustrates the range of preferred analyses I have encountered and the source(s) of such advice. Some of these analyses require fieldwork; some do not. For any one project, not all of the issues listed in Table 4-3 need to be examined.

The involvement of competent social scientists who can interface with environmental managers on these issues is crucial in developing effective public participation and reports of high quality. If these special-subject areas are competently managed and if the synthesis is well handled, the conclusions and recommendations of the study have a low probability of

Table 4-3. Range of social science analyses relevant to ecoplanning.

Amenity and disamenity features both natural and human-made

Boundary definitions of the "community" or "neighborhood" affected

Special lifestyles that influence land use

Archaeological sites

Historic sites (cemeteries, battlegrounds, churches, etc.)

Heritage houses and buildings

Community icons (symbolic areas or features)

Differential economic impact of a project or new technology on socioeconomic groups

Community health profile

Epidemiological problems resulting from toxic waste

Tangible and intangible economic values

Community and neighborhood disruption

Historical patterns of land settlement and land use

Political orientation or ethnic composition of target socioeconomic groups to be affected by a proposed development

Socioeconomic composition and demographic structure of a community linked to recreation demand and facility use

Buffer zones for incompatible land uses

Unresolved issues from earlier involvement with a client that can impede dialogue and mutual trust error and a higher potential for implementation. If there are "hidden agendas," such as a public land purchase to deliberately expropriate landowners of the "wrong political party," a social scientist is more likely to identify such a situation than a technical person, who is more inclined to take social situations at face value.

Critical Path Analysis and Achieving Integration of Disciplines

It is one thing to reach into specialized areas of knowledge and quite another to sort out significant elements from those that are not and then develop relevant time-study relationships. It is common for inexperienced specialists to become buried in technical material and to waste large amounts of time processing and writing material that then requires heavy editing. Since time of the specialists and the project coordinator or environmental manager must be paid for by the client or be absorbed by the firm, the client's management staff quickly perceives if irrelevant work is being done or if deadlines are being consistently missed. A client is not likely to repeat this experience with a firm.

Even though there is no substitute for experience, some general comments on avoiding information overload and cost overruns, that is, to prevent slippage in staff time, might prove helpful:

1. To cover at the proposal stage the various technical areas, the terms of reference should be discussed by the prime consultant with the cooperating specialists. This discussion includes whether or not the terms of reference are appropriate and realistic. Problems of scale of detail, the time frame, and the seasonal nature of natural phenomena need attention. This tailors the proposal so that wasteful "fishing expeditions" for irrelevant data do not jeopardize acceptance of the proposal. A well-designed proposal sets the stage for effective collaboration.
2. To avoid confrontations, it is important that the organizational and decision-making structure of the group be understood by all the specialists. In a hierarchial structure, decisions are more direct and timetables can be met, but the solutions tend to be less creative. Ad hoc structures may be more creative but may have trouble meeting deadlines. A hybrid organizational format, similar to a wheel, can be envisioned: specialists on the rim connected to a leader or senior cadre at the hub and to each other along the rim. Such a structure retains control, allows for creativity, and facilitates meeting deadlines.
3. To bring a specialist into the study group, especially if he or she has not worked with the other members of the team, requires that the specialist understand the relevance of his or her effort to the total study, the time frame for the work, and the meshing of the various separate contributions of the team.

4. To meet the final deadline and to ensure that there are no incomplete aspects in the study, at the midpoint the leader or core team should review the work of all the specialists at a general staff meeting. A review allows all specialists to present and to listen; new ideas, often of significance, can emerge unexpectedly. It also identifies any fine-tuning or corrections to the efforts of the specialists and any inconsistencies or disagreements; later it simplifies the report writing.
5. To facilitate the report writing the communications specialist or editor of the final document should attend all staff meetings. This is particularly important if a professional editor is being hired.

Utilizing these five steps, quality environmental work of an interdisciplinary nature is accomplished. Human impact over time can be perceived, biotic potential or landscape capability identified, biotic thresholds and toxic conditions determined, and land management procedures recommended in a dynamic and innovative framework.

A critical path schema or logic diagram depicting the time and linkages between specialized areas of knowledge and the integration of the environmental component in a master planning and implementation process for a proposed hospital development is depicted in Fig. 4-1, originally prepared by Vincent Moore (personal communication). This plan would dovetail with similar critical path networks for architectural, landscape architectural, and engineering services. Another, similar example, which does not have a time component, is presented by way of contrast: Fig. 4-2 was developed for shoreline planning purposes (Fisheries and Oceans 1981).

If an outside group attacks the study, the attack is usually directed toward a specialized area, such as in the social or natural sciences, by a specialist in that discipline. If an individual or group can discredit one segment of the study, it can raise doubts as to the validity of the entire study. If the best available technical advice has been obtained, and if it has been integrated with thoroughness, the project coordinator should be able to discuss credibly any technical issues raised. Credibility in an environmental study is enhanced by broadening the scope of the component studies and by the skill exhibited in integrating the separate discipline inputs. This generally requires "touching base" with locally recognized key people in the area. A study can just as easily be jeopardized by inattention to the matters of discipline integration, which have been raised in this section.

The Role of the Communications Specialist

As alluded to in the preceding section and noted in Fig. 1-3, the ability to understand the published technical information on ecosystem structure and function generated by a wide variety of professionals is difficult.

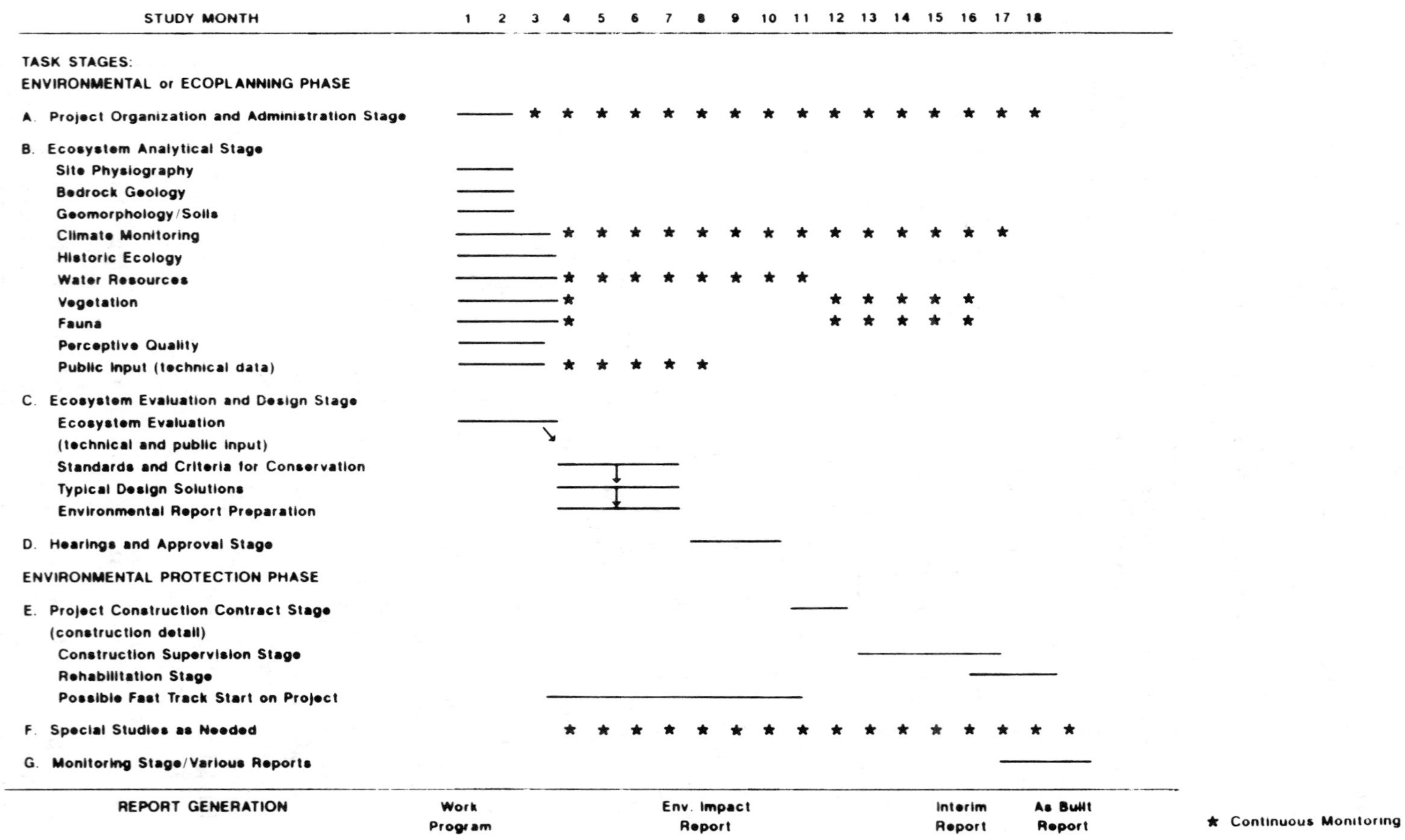

Figure 4-1. The environmental component for an idealized 18-month environmental program for a large scale project.

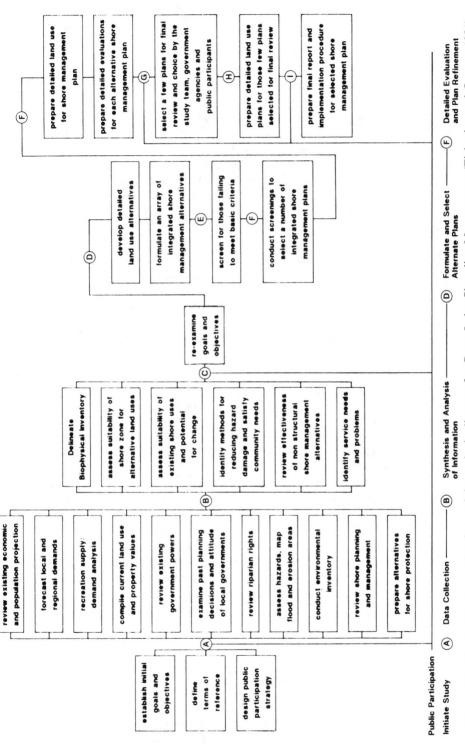

Figure 4-2. General land use planning activity or logic diagram for Great Lakes Shoreline Management (Fisheries and Oceans 1981).

105

Specialized language and concepts, often camouflaged by seeming to be "common words," and the need to condense a report while avoiding the introduction of technical errors into the text are major obstacles to accurate and clear communication. I have found that whenever more than two or three specialists of varying disciplines are involved in a study, language or jargon often blocks in-house integration of the data and communication of findings to the client. The introduction of an individual trained in writing, not necessarily having a specialized technical background, facilitates the production of a quality report. I have called such a person a *communications specialist* (Dorney 1973), a term somewhat broader than the usual definition of an editor.

The utilization of such a person can be made efficient by careful sequencing of the use of such a person's talents. If the communications specialist is brought in when the final report is being done, it proves difficult for this person to begin to work effectively, since it takes up to a week to get a focus on the project. This focusing period means that deadlines are not met. A communications specialist should attend key meetings and meet privately with individual consultants who are preparing their part of the report (see Fig. 4-2). In addition, an opportunity to meet with the client can be useful to pitch the report properly. A brief visit to the study area helps the communications specialist visualize the issues and become familiar with the site. When questions of terminology arise, the communications specialist can call on a technical specialist to resolve issues without requiring intervention by senior staff. Prepared in this way, the overall quality of a report is much higher, as it will be coherent from a stylistic and technical viewpoint. Because the use of a communications specialist increases the cost of the report by about 15 to 25 percent, it needs to be budgeted at the outset.

I have seen many excellent technical reports in environmental management fail to achieve any credibility simply because editorial control of the text and graphics was not appropriate to the client's needs or to the general readership. Although such communications specialists are hard to find, the additional cost of bringing them into a final report process can be justified, particularly on sensitive projects. Many competent part-time people are available for this work, making it unnecessary to have a full-time person on staff if the work load is irregular.

Regional Aspects of Environmental Management Practice

Experience from reading reports prepared by environmental consultants shows that an ecological group working in a regional area, say within a 150- to 300-mile or 300- to 500-km radius, develops greater technical skills and a profound knowledge of regional biotic and ecosystem variations and of key studies. An outside environmental specialist, or hired gun, lacks both local field experience and the detailed knowledge of local experts in fields such as archaeology, historical ecology, and soil science. Also, an outsider lacks contacts in organizations and institutions that have important archival material, especially theses. Because of the complexities of ecosystems and the vast amount of technical information that may be available relating to them, outsiders may have a problem in credibility if cross-examined by lawyers at hearings. This occurs particularly when legal intervenors are backed by local, well-qualified specialists. It is my belief that an environmental manager should not work outside his or her own region without requesting assistance from local environmental or scientific specialists. However, recruiting competent local people to develop a mixed team and organizing them effectively are time-consuming. Since either the client has to pay for time spent or the professional must donate time, working outside of a firm's own region can be less than fruitful. An example is appropriate here.

Some years ago I was asked to examine a 500-acre (202.42-hectare) institutional construction site and to comment on the environmental impact of the conceptual design. With only fragmentary soil maps and little understanding of the local forest ecology-physiography-climate relationships, I could only raise some serious doubts as to the survival of some beautiful, well-stocked, but mature hemlock stands. By suggesting the involvement of a local forester and local soil scientist, I not only protected myself but increased the probability that a correct impact interpretation would be made. When a local forester looked at the hemlock stands the next day, he agreed that these stands would not survive construction impact. A more thorough forestry and physiographic study conducted over a 6-month period by an interdisciplinary team under our guidance verified these conclusions and altered the design concept. To have tried to work without any local forestry-soils input would have been most risky professionally, and unnecessarily costly to the client if the conclusions had been incorrect.

Put another way, I see a clear distinction between personnel who are regionally competent and a project director who needs not be to the same extent. The latter person has generic organizational skills not tied to a specific region. The self-contained environmental management firm analyzing environmental problems from the arctic to the equator, from desert to oceans areas, is not workable technically unless it has a large, wide-ranging managerial staff and a large reservoir of local talent on tap. If quality and credibility are to be achieved, some local competent full-time staff have to be employed. Whether or not such regional firms remain independent or prefer to have business ties with colleagues or other regions is, perhaps, not important. What is important is that we not see the instant expert arriving by plane, waving a magic wand over the area in question, and pontificating about land use and environmental quality issues. Liability claims clean up shoddy work, and hearing boards grind out the truth, but these are painful ways to discourage the suitcase experts.*

A further necessity in providing high-quality professional environmental management advice is the need to understand the larger spatial context in which a particular study area exists. For example, a boreal ericaceous bog, common in northern Wisconsin, is ranked low for preservation; in southern Wisconsin, because of the abundance of calcareous tills, such a bog is rare and so ranks high for preservation. Further, a regional-planning concept, such as the interstate zoning of estuarine areas for preservation and management along the Georgia and Florida coast, can be reinforced by integration of local site plans within this larger region.

In my experience, the client and his consultants often do not take the time to learn, or are not familiar with the myriad of changes and proposals in land use management or policy going on around them. Coming in as an outsider and becoming entangled with these issues during or at the end of a project is wasteful and embarrassing. Avoiding these entanglements requires a background of work and local ties with various planning groups, resource management agencies, and public organizations.

Besides hiring local qualified scientific specialists sensitive to such regional nuances, the client or agency also can hire an expert panel to oversee the ecological design work at the various stages. Such an expert panel may include some local people, individuals from relevant regulatory agencies, key politicians, and knowledgeable outsiders. This concept was applied, for example, for the Amelia Island Study by the Sea Pines Company (Wallace et al. 1972).

The Environmental Impact Statement as It Relates to Environmental Practice

The evolution and institutionalization of an environmental impact statement into professional practice and into legislative requirements in many states, in Canada at the federal level, and in some provinces have fostered substantial interest in the predictive use of environmental management for assisting decision makers before specific facilities are built. Some agencies prefer the term "environmental assessment" or "environmental report," because the word "impact" suggests that damage or "insult" will be done to an ecosystem. Whichever term is used, the environmental assessment concept has resulted in demand for more ecosystem analysis prior to undertaking specific courses of action.

---

*By way of analogy in the medical and veterinary professions, regional patterns of disease exist, making it necessary for physicians and veterinarians to spend some months or years becoming acquainted with the nuances of an area. Similarly, a lawyer must know local statutes, decisions of the local courts, and local judges' rulings.

The conceptual linkages and operational approaches between environmental assessment, land use zoning, and official plans offer a fertile ground for expanding assessment of a facility to a larger assessment of an official plan or a policy plan (Dorney 1985). In any case, the institutionalization of environmental assessment laws in many jurisdictions has stimulated the establishment of environmental firms and generated significant cash flow for interdisciplinary studies. Thus the importance of environmental assessment in organizing and developing a private practice cannot be understated.

In my judgment, by broadening the term "environment" to include not just biophysical impacts but also social impact (presented in Chapter 3), the environmental impact assessment (EIA) affects a much broader public group, which can contribute to such a study. Furthermore, doing environmental assessment provides political and media visibility to ensure that thorough analysis and public discussion take place.

To ensure that all the physical, biological, and social processes are given proper consideration, a number of matrices have been proposed (Leopold et al. 1971; Fischer and Davies 1973). As a professional practitioner, I see such matrices as a useful approach for government agencies and administrative personnel lacking depth in environmental sciences and resource management. Matrices organize terms of reference for ecological studies, point out potential interactions, and direct specific questions to various consultants when they are interviewed for contractual work. Furthermore, matrices help ensure that no loopholes have been missed in the study design.

However, matrices are unable to provide quantitative answers to technical and judgmental questions and if used for these purposes can damage the effectiveness of a study. An alert professional of almost any discipline can quickly discredit a matrix, especially when weighted values are summed to indicate preferential solutions. Unfortunately, ecosystem sturcture, function, and sensitivity cannot be deduced by matrix analysis alone. Numerical scores tend to gloss over the systems aspects of the analysis. It might be argued that such a numerical analysis is better than the judgment of an environmental manager familiar with a regional environment, but if it is an either-or situation, I prefer the latter. Ideally the two might be done concurrently so that broad judgment that is regionally based is not automatically ignored. Sensitivity analysis can help if scoring is done; this approach, by varying the numbers, can determine the critical factors that are tipping the solution in one direction or the other.

For example, in considering three alternative designs for a new Pickering International (Toronto) Airport, identification of the severity of impact of each development on surface water quantity and quality does not, in itself, provide the only useful perspective. Factors such as the renovation potential of an existing stream, the secondary effects of development on this and adjacent water bodies, and their ability to tolerate change if a flood protection dam is built (blocking salmon movement or warming one reach of the stream) can be equally important.

If small projects could be undertaken side by side using these matrices and numerical analysis techniques and then coordinated with interdisciplinary team approach for determining ecosystem dynamics—the method I prefer—the matrix methods are better utilized. The greatest danger in the matrix method is that it is often prepared by a person from a single-discipline background who feels compelled to fill in every blank space, bypassing years of professional expertise available in the region which can provide a better overview of the system properties in question.

The 51 technical principles outlined in Chapter 2 provide a conceptual approach to environmental management that could be used in conjunction with impact matrices and other evaluation procedures. The two approaches reinforce each other, avoiding the embarrassment of missing some important issues. For example, principle 1 deals with identification of the decision making or planning process to be followed. If an ad hoc process is being used, the impact assessment may require continual alteration and modification, jeopardizing its quality. Other philosophical and ethical issues raised by environmental assessment can be equally troublesome unless appreciated before such an impact assessment begins (Dorney 1976).

Although the concept of an environmental impact assessment has been a useful device for encouraging predictive ecological studies, it has the disadvantage, as suggested above, that once done and approvals given, the environmental manager makes no further contribution. The technical deficiencies from the perspective of an ecological scientist are also worth considering; these aspects are summarized in two recent publications (Rosenberg et al. 1981; Beanlands and Duinker 1983).

Unless environmental assessment becomes part of an overall environmental management process—which considers all stages of project planning, design, hearings, and environmental protection as part of implementation—the results are less than satisfactory. The EIA document can keep the approval agencies happy, but little substantive change in the planning, design, and the implementation process has been achieved.

Particularly troublesome is dealing with a problem of uncertainty that may require experimentation and adaptation, such as an ecosystem remedial action program or rehabilitation program for a harbor with toxic sediments. In the face of uncertainty as to what may happen, being free to take immediate corrective action in response to continuous monitoring is crucial (Dorney 1977). The environmental assessment approval procedure is inherently fixed, not adaptive; it works well where certainty is expected, but works poorly in the reverse situation.

Bibliography

Apostel L et al. (1972) Interdisciplinarity: Problems of Teaching and Research at Universities. Paris: OECD.

Beanlands GE, Duinker PN (1983) An Ecological Framework for Environmental Impact Assessment in Canada. Halifax: Institute for Resource and Environmental Studies, Dalhousie University.

Dorney RS (1973) Role of ecologists as consultants in urban planning and design. Hum Ecol 1(3):183–200.

Dorney RS (1976) Environmental assessment: The ecological dimension. J Am-Water Works Assoc 69(4):182-185.

Dorney RS (1977) Hindsight evaluation of environmental assessments, a review. In: Environmental Impact Assessment in Canada: Processes and Approaches. Toronto: Institute for Environment Studies, University of Toronto, pp 181–183.

Dorney RS (1985) Predicting environmental impacts of land-use development projects. In: Whitney JBR, Maclaren VM (eds) Environmental Impact Assessment: The Canadian Experience. Environment Monograph 5. Toronto: University of Toronto, pp 135–149.

Fisheries and Oceans (1981) Great Lakes Shore Management Guide. Ottawa and Toronto: Environment Canada and Ministry of Natural Resources.

Fischer DW, Davies SG (1973) An approach to assessing environmental impacts. J Environ Mgmt 1:207–227.

Leopold LB, Clarke FE, Hanshaw BB, Balsey JR (1971) A procedure for evaluating environmental impact. Washington DC: Geol Surv Circ 645.

Rosenberg DM, Resh VH, et al. (1981) Recent trends in environmental impact assessment. Can J Fish Aquatic Sci 38:591–624.

Wallace, McHarg, Roberts, Todd (1972) A Report in the Master Planning Process for a New Recreational Community: Amelia Island, Florida. Philadelphia: Sea Pines.

5
Operation of an Environmental Management Practice

This chapter explores various aspects of the relationship between clients and the professional environmental manager. The issues involve contacting prospective clients, preparing proposals, organizing a team effort, determining the appropriate scale of detail, the timing and seasonal aspects of field surveys, phasing of the work, ensuring confidentiality and credibility, preparing for public or judicial hearings, establishing fees, and determining cash flow. Understanding and managing these various elements make a practice successful professionally and financially and influence the performance and credibility of the firm and its partners or associates. The matter of professional ethics (discussed in Chapter 2) will be touched upon briefly here as well.

Many issues regarding how to handle clients are common to many other professions. Nonetheless, because of certain peculiarities of an interdisciplinary organization, some issues are unique to an environmental management practice. Furthermore, certain technical matters are equally distinctive, justifying discussion in this chapter.

Contacting Clients

Foremost in contacting clients is an understanding of the clients' level of knowledge about the range of issues to be addressed. If clients are unaware, you may have to educate them, working at all levels of the organization from the bottom branch or junior level to the top echelons. Sometimes circumstances can conveniently give your point of view a "hand-up," such as receiving media attention; call-backs can drive the point home, and work can quickly follow.

Second is the importance of marketing surveys. If available, or if they can be commissioned, survey results target a group of clients who need services of the kind you offer. Next is the direct approach to clients. "Direct" means the preparation of lists of prospective government agencies, competitive and collaborative private consultants, and private corporations. Writing or calling them to describe your group's qualifications gets you onto their desk. Appointments over coffee or lunch reinforce the business card, notices in professional engineering journals and newsletters, introductory letter, or brochure.

An attractive brochure introduces the principals and the team to the client. It demonstrates the scope of services offered and the projects completed. Work done individually or done before incorporation or partnership should be differentiated from that done as a full operating team. Persons shown on the brochure as "associated" with the firm, that is, part time, also should be differentiated from those who are full time. If associated people are well-known individuals, this helps to build overall interdisciplinary strength and reputation without requiring the overhead costs their full-time services would demand. However, advertising professional services is always done in a low-key manner; any other forms are in poor taste and could alienate other professionals, potential clients, and the public.

Using a direct-contact approach allows the young professional at least to get into the action. What develops from this approach is up to the skill and persuasiveness of the professional and, to some extent, luck. After three or four years, quality work will speak for itself, but the survival of a new firm to this point depends on consideration of these details.

A number of indirect approaches to clients are equally productive and should be mentioned, as they may appear at first glance to be peripheral to obtaining commissions or contracts. First is the potential value in participating as a member in trade organizations, professional organizations, research groups, service clubs, government advisory committees, and foundations. Such groups, which rarely fund work directly, do provide many formal or informal business contacts and provide access to inaccessible information and to key individuals. These groups welcome assistance on specific projects, especially when funds to finance such projects require volunteer professional labor. Working on briefs, committee reports, and program organization allows the professional to visit other offices and meet people—make contacts. If one is alert to prospective jobs, such as budget tag ends to be used up before the end of the fiscal year or peak work loads caused by personnel who have resigned unexpectedly, these situations can be an opening wedge for occasional or even steady contract work.

The second indirect, but less obvious, way to contact clients is by presenting talks and technical papers and by publishing research or case studies in journals that reach groups using environmental management services. In this regard it is important to note that "preaching to the converted" also means preaching to competitors. "Preaching to the heathens" is more demanding, but it is more likely that prospective clients will hear the message.

A third indirect method is to attend workshops where the prime motivation in attending is not to learn, as the subject may already be well known or it may be elementary, but to meet those participants who have come to learn. Considerable time is available over coffee and meals and in the hotel after the sessions to meet prospective clients in an informal setting. Contacts made at such seminars can open doors later.

Since the interdisciplinary team is a requisite for an environmental practice, setting it apart from most legal and many engineering or design practices, this organizational difference should be pointed out to clients. Clients can better appreciate that they are buying both the services of skilled principals and the combined expertise of a team having substantial discipline depth as well as transdisciplinary breadth. Promotional time easily exceeds 5 percent of the total work load of a firm; 15 percent is probably a more reasonable figure. Ironically, when you are the busiest, it is all the more important to remember not to relax promotional efforts. Peaks too quickly become troughs; being in a trough is too late for promotional efforts to quickly boost billable time and morale.

Preparing Proposals

In some cases—in my experience about a third—a developer, government agency, or engineering/planning firm does not perceive that it has an environmental quality issue until the problem suddenly arises. In these circumstances the client puts the environmental manager under severe time constraints to provide some credible answer and to bypass the preparation of a formal work program and budget. On the other hand, if the client has some lead time, terms of reference are prepared, and requests for proposals are sent to a list of qualified consultants. At whatever point the environmental manager is brought into the problem, there are crucial aspects of problem definition, scale of effort, and time required or justified that need to be thoroughly explored before a proposal can be prepared.

It is not uncommon for the terms of reference to provide only a partial explanation for requesting environmental advice. A job may have touchy interagency or political overtones. A personal meeting with the prospective client can probe or uncover the actual rationale for requesting environmental advice other than that explicitly put in the proposal. Acceptance of the proposal usually depends not only on showing clients how to answer their explicit problem, but on identifying the implicit problem itself in a sensitive manner.

Any previous work or work in progress of a similar nature is vital in developing a feel for the problem before preparing a proposal. Following this, a site visit—*of critical importance* in almost every case—to the proposed area, which may take one-half to a full day, familiarizes one with the lay of the land. Visiting similar sites that have been studied, approved, and developed is useful. Clues to environmental issues, such as leaf damage by air pollutants, tip dieback of sensitive species, and various types of former or present land uses as evidenced by old barns, silos, field successions, abandoned orchards, and fence row patterns, should be noted and discussed with any other consultants or subconsultants. Prepared from this perspective, the proposal shows both general thoroughness and a specific depth of understanding of likely key issues.

Based on the regional background and experience of the environmental manager and his or her specialized contacts in other disciplines, it should be possible to prepare a coherent proposal. The proposal includes a short statement of the issues needing analysis, the time required, the personnel involved and their daily rates, the company markup, the total estimated cost, and the delegation of responsibility for job management and final report preparation and any hearings.

Proposals may require only one or two hours preparation for a small job but up to a week for a large complex project, especially if there is competition with other firms, requiring enlarging the team to include other subconsultants. If the terms of reference miss key issues, identifying them for the client helps establish the firm's credibility; such an action may set you apart from the competitors.

For a proposal format, using word processors facilitates preparation of new proposals by "filling in the blanks," so to speak. If junior personnel assist in preparing proposals, this formating saves considerable time. Usually, time spent on preparing proposals is not recoverable: if the client accepts the proposal, some time spent can be partially recovered, because the work moves along much faster. Occasionally, for complicated proposals, agencies may allow funds for travel, telephone tolls, and secretarial and duplicating costs; less commonly, they allow recovery of all of the staff time spent or pay a flat fee for preparation.

On occasion, government agencies use proposals to gain knowledge from consultants on how to do the work in house; then the agencies hire graduate students, at a reduced salary, to complete the work either under contract or in house. In addition to the question of the quality of the resulting work, unethical as this may be, the word quickly gets around among the consultants; agencies doing this too often "find the well dry."

Another trick agencies have been known to play is to send requests for proposals to as many as 15 consultants. If this is the case, the probability is so low of any one group getting the job that it is hard to justify spending a great deal of time on such proposals. When a firm is short listed (competition reduced to three or four), considerable effort is justified on the increased probability of one-third or one-fourth. On a recent occasion, 15 firms had been contacted. I suspected the agency apparently had already "picked" the company to do the work, because one of the companies on the list was a new firm headed by a man who had recently left the agency's service. Before leaving the agency he had written the terms of reference for the study. The requests to the 14 other firms seemed to be a ruse to legitimize the selection. When our "network" told us what was behind the proposal, we spent little time preparing it. The work was awarded as I had expected.

Care in problem identification when preparing the proposal is essential if quality work is to be done. It is unusual for a client to agree to substantially change the terms of reference and the budget once the proposal is accepted and the project is under way, unless unforeseen circumstances have arisen. Such changes generally require additional meetings, strong justification, and painful adjustments in a tight timetable. When the proposal is presented it is useful to identify any uncertainties so that, as the work proceeds and requests for additional effort or time are made, such requests do not come as a complete surprise to the client. If the project contains a certain amount of unknowns, a trial sampling to establish degree of effort, and thereby cost, is not unreasonable. Depending on the outcome of this trial, firm costs can be negotiated to the advantage of both parties.

Budget overruns, which are a continual management headache, result from poor initial problem definition and weak job management. For this reason, senior staff whose financial future is tied closely to the firm should be involved in problem definition and budget preparation to reduce the probability of being hurt financially when a contract is signed and then difficulties arises. On a large project, it is often useful for two senior members of a firm to independently prepare a budget and then compare their estimates; this reduces the probability of error.

During budget preparation on a proposal as a subconsultant with a prime consultant, it is critical to know the minimum time and cost necessary for quality work. Some costs are more or less proportional to the acreage to be studied (if it is a land use problem) or linear miles of facility (if it is a pipeline or highway) and are inflexible. Others, such as social surveys, can be adjusted if the client is willing to use less expensive methodologies or if other consultants on the team can cover the work adequately. We have found it necessary on more than one occasion to withdraw from a project awarded to a consortium of which we initially were a part, because the budget after approval was reallocated internally to such an extent that quality fieldwork would have been impossible. A firm approach to the prime consultant concerning this potential pressure on budget is vital, whether the project is at the proposal stage or after the award has been made. Retaining long-term credibility can lead to some painful short-term cash flow problems if, after the consortium is funded, your withdrawal is necessary.

The unsolicited proposal is another interesting consideration. In this circumstance a new issue has emerged which you believe can be studied to the advantage of the agency or corporation. Even if unsuccessful, this unsolicited proposal may trigger action at a later date; your name is then attached to it, so to speak. Another indirect advantage to a firm, if an unsolicited proposal is accepted, is that you are the first through the gate.

The tag ends of an agency's budget year can also present opportunities for work. Frequent in-house contact is necessary, preproposals need to be prepared, and key individuals on the committees or whoever is holding sole authorization to proceed need to be familiar with your group.

To quote as a subcontractor to more than one prime contractor on the same proposal is possible if each prime contractor agrees. We then bid the same amount to each; we do not take part in proposal strategy sessions, as this could compromise either of the prime consultants' approach or their competitive budgets. This last point is important, because collusion on bids is illegal; the perception of collusion can discredit a firm.

Some problems may not be amenable to environmental analysis because of time and budget constraints or the nature of the situation. It is quite common for clients to want quick one-hour or half-day reconnaissance-type fieldwork or service by telephone, without allowing some time to determine the complexity of the issue and the future difficulties of liability and court testimony this may entail. This situation also requires a firm hand and the ability to say "no" tactfully, pointing out the danger of short cuts to both the consultant and the client.

On a few occasions, consultants are brought in as "hired guns." A hired gun is an outside consultant who springs the bad news on a client, another consultant, or a rival agency so that a particular point of view is heard. Even though one's ego may be reinforced by being a hired gun on the "winning team," the hazards from losing are equally apparent. Such situations, if set up to solve personality or interagency conflicts, become professional confrontations; they are not for the faint of heart. Sometimes, however, if one wins, proposals and work can ensue from the camaraderie that results from a successful "raid" on another's "territory."

Organizing the Team Effort and Project Management

In structuring a consulting company, the principals are the key to its success in an operational sense as well as in a financial sense. There are a number of ways of achieving both. First is to spread the management load by the delegation of some aspects of project management to middle-level staff. This is cost-effective and builds the overall competence of the firm. Tight control is required to avoid budget overruns, lack of follow-through, problems of meeting deadlines, etc. Second is providing a time/cost breakdown for each project; this is especially critical where concurrent projects are under way. Third is to call frequent team meetings to review progress on studies to increase overall project coordination, effectiveness, and efficiency. A more subtle spinoff of such meetings is the cross-discipline integration that emerges at such staff meetings, especially if the project manager encourages the group to relax and "open up"; however, he or she must also ensure that discussion stays on topic.

Report editing requires special skills. The communications specialist (discussed in Chapters 4 and 6) must be able to understand technical language and translate it into a more understandable common usage without distortion. If the communications specialist attends team meetings, this helps develop both a feel for the project and a feel for the language used by the specialists involved in the study. If a critical deadline must be met, a partially edited draft, clearly stamped "DRAFT," can often be submitted. It should quickly be edited and vetted by the team as to its accuracy and content.

When the final report is edited and essentially finished and the conclusions and recommendations have been prepared, feedback to all subconsultants and involved individuals prior to submission to the client ensures that no points have been omitted or misinterpretations conveyed. If any member of the team disagrees with the recommendations, the content of the report, or the editorial changes required to achieve clarity of expression, the project manager must have sufficient time to iron out the disagreements before sending the final report. Having this team feedback can ensure that any subconsultants who may be asked to testify before hearing boards will support the study conclusions or recommendations without reservation, perform well as witnesses, and show interest in doing further work with the firm.

Determining the Appropriate Methodology and Scale of Detail

A successful project requires some understanding of what constitutes the correct relationship between scale of detail for fieldwork in the social, natural, and physical sciences and the environmental problem to be analyzed. Although there are no absolutes, some guidelines can be discussed.

For social analysis, the spatial relationships may be crucial. If community or neighborhood impact is being addressed, the boundaries of the community or neighborhood exist in people's "mental maps" as well as in fixed political lines. These realities may be quite different; confusing results then emerge, and time is wasted. For example, if nomadic people are present, like some Inuit groups in the arctic, their numbers and

Table 5-1. Scale of analytical detail useful to ensure appropriate integration with other design and engineering disciplines.[a]

Project description	Scale of detail
Strategic studies or policy planning for regional purposes (2,000 to 10,000 km²/1,000 to 5,000 sq mile area)	1:25,000 to 1:250,000 maps. Contour interval 5 to 50 meters/10 to 100 feet
Feasibility planning for a major freeway corridor (10 to 100 km/6 to 60 miles)	1:10,000 to 1:50,000, or 1 inch to the mile (1:63,360). Contour interval 2 or 5 meters/5 or 25 feet.
Master plan or conceptual design for a new town or a major park (500 to 10,000 ha/1,000 to 20,000 acres)	1:5,000 to 1:10,000 maps, or 1 inch to 400 feet or 1 inch to 800 feet.
Secondary Design (400 to 2,000 ha/ 1,000 to 5,000 acres)	1:5,000 maps, or 1 inch to 400 feet
Detailed facility design for lot layout and individual tree management (50 to 500 ha/100 to 1,000 acrea)	1:500 to 1:1,000 maps, or 1 inch to 100 feet. Contour interval 1 meter or $2\frac{1}{2}$ or 5 feet with interpolation to $2\frac{1}{2}$ feet

[a] See also Fig. 3-4.

locations may not be found in archival, legal, and administrative material, but they know where their territory is.

The scale of maps selected depends on what base mapping is available and what final mapping is required. Examples from various projects have been discussed in Chapter 3 (see Table 3-1 and Fig. 3-1) and are expanded on in Table 5-1. Inappropriate base mapping leads to greater resource mapping expenses if the base map needs to be enlarged photographically more than twice. Generally a two-times shift in scale is as far as a map can be altered accurately. The scale for all separate resources depicted on maps should be the same as well, as this facilitates final evaluation, especially if an overlay mapping system is used.

In addition to these general guidelines of the appropriate scale to use, some particular issues of methodology and scale relate to the separate resources inventoried and are elaborated upon in the following section. For additional reference material along these lines, Lang and Armour (1980) should be consulted.

Soil and Topographic Surveys

Soil surveys for agricultural purposes are generally done in the field at 1 inch to 1,320 feet, or 1:20,000, using aerial photos, and then published at 1 inch to the mile, or 1:50,000. For regional planning, new town planning at the conceptual scale, and large provincial or state park conceptual planning, this scale presents no problem if the maps are reasonably recent. To expand such maps for the purpose of detailed subdivision conceptual design using engineering maps prepared at a scale of 1 inch to 400 feet, or 1 : 5,000, and a contour interval of $2\frac{1}{2}$ feet or 1 meter, unacceptable error by distortion and by exclusion of real features is introduced.

An example from a study west of Toronto (Dorney 1973) illustrates the point. At an actual mapping scale of 1 inch to 1,320 feet reduced and published at a scale of 1 inch to the mile (1:63,360), the crude outline of the valleys is visible (Fig. 5-1). For the identical area remapped at a scale of 1 inch to 400 feet without adjustment in scale, an additional valley complex emerges in the center (Fig. 5-2); an important geomorphological unstable soil formation (Brockport) associated with this valley wall also emerges. Utilizing the original soil map published at 1 inch to the mile (Fig. 5-1) for road and lot design is clearly misleading and inappropriate. Using it could lead to errors of omission or commission and hence expose the consultant to a liability suit.

Unfortunately, there are time and cost constraints for transforming geomorphological-soils maps from 1 inch to 1,320 feet (or 1:20,000) to 1 inch to 400 feet (or 1:5,000). Soil survey work is best done in spring after the frost is out of the ground; it is virtually impossible to do in winter. A soil scientist knowledgeable in regional soil and geological formations can map onto a 1 inch to 400 feet (1:5,000) or 1 inch to 200 feet (1:2,500) topographic map about 25 to 50 acres (10 to 20 hectares) per day. Up-to-date stereo aerial photos and new engineering survey topographic maps with 5 foot or 2 meter contours are needed for the work. A total cost can be calculated based on the daily rate of the soil scientist.

Topographic mapping is important at all scales of detail. When subdivision level planning or site planning for a campground, where road network, individual camp sites, and washroom facilities are to be located is undertaken, slope constraints to development become important from a transportation safety point of view and from a storm and sanitary sewer standpoint. Generally, slope constraints are needed at $2\frac{1}{2}$- to 5-foot or 1-meter contour intervals. On the other hand, for conceptual design purposes for a major urban area or for a feasibility study of a highway corridor (1:25,000 or 1:10,000 scale mapping), 25-foot or 5- or 10-meter intervals would be sufficient.

Vegetation Mapping Surveys

Vegetation of various kinds including forest stands can be mapped and analyzed in several ways. If the issue is one of a regional planning and zoning at a scale of 1:50,000, a simple classification of vegetation types

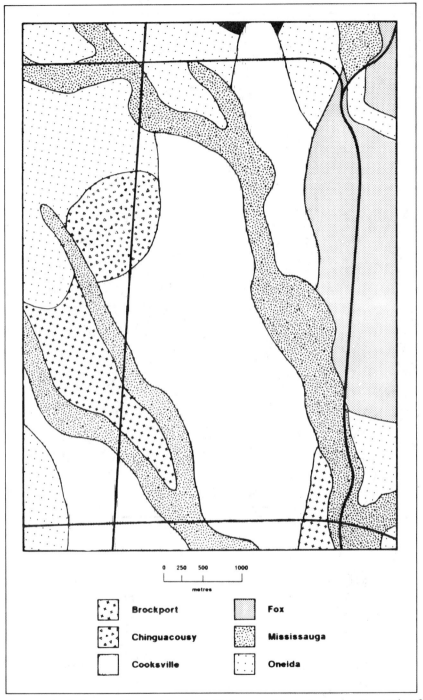

Figure 5-1. Soil map prepared in the field from aerial photographs at a scale of 1:15,840.

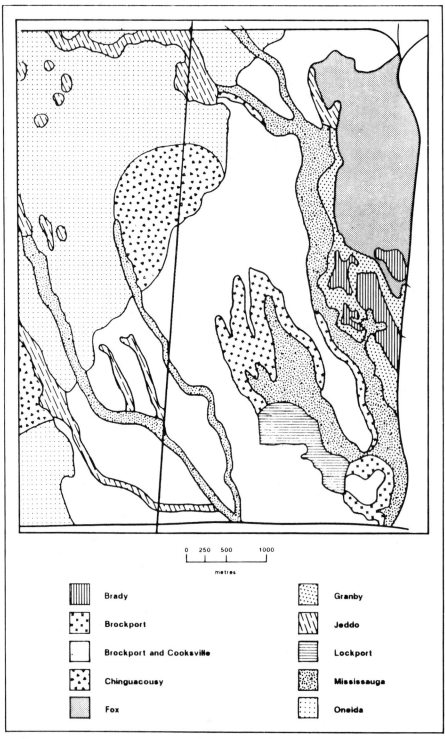

Figure 5-2. Soil map prepared in the field from aerial photographs reinterpreted for a scale of 1:4,800 but displayed at the same scale as Fig. 5-1. Note the Brockport soil series (see page 120 for explanation).

into old fields being invaded by forests, sand dunes, back dunes, upland-lowland-natural forest areas, and plantations may be appropriate; major plant community types are identified with some indication of dominant species. If the issue is economic forest production, conventional forestry maps showing species, stocking, height, etc. are needed. If the focus is on environmental sensitive areas (unique natural areas), special biological attributes such as rare or endangered species need indentification.

For new towns and state or provincial park conceptual planning, qualitative rather than quantitative description of each forest stand often is sufficient at a scale between 1:2,500 and 1:20,000, or between 1 inch to 200 feet to 1 inch to 1,320 feet. Limited quantitative analysis may be selectively done on those forest areas likely to be particularly affected or disturbed. A qualitative land unit biotic evaluation scheme I have found to be useful to engineers and designers (Table 5-2) ranks each vegetation unit by using a three-tier description-function-value scheme: *description* of the vegetation includes 14 parameters, such as sensitivity to human impact; *functions* include 10 biophysical processes the vegetation may perform; and 12 potential *values* are identified (Dorney 1977). This analytical approach is designed for environmental planning, assessment, construction supervision, and monitoring in both urban and urbanizing areas; it is justified where land values are high and delays in approvals could be costly.

For quantitative plant ecology the point-centered quarter method is the preferred one, although transects, quadrats, or visual evaluation methods (Dorney and Leitner 1985) also are appropriate. Quantitative data can be combined, of course, with qualitative evaluation techniques. For example, rare plant species, specimen trees, or unusual plant communities as small as 1 to 3 acres (1 hectare more or less) in size can usually be shown on a 1:5,000 scale maps or on 1-inch to 200-00 to 1-inch to 1,320-foot maps and added to the survey information. An example of information prepared in this manner is shown in Table 5-3 and Fig. 5-3. This mixture of quantitative and qualitative information allows the designer to select suitable parkland or parkland units, identify corridor and servicing impacts, and preserve maximum natural vegetative cover capable of surviving urbanization.

At a more refined site design scale (1:500), planning for tree preservation is a sequence from site analysis (Fig. 5-4) to road and lot layout (Fig. 5-5), to environmentally sensitive site design (Fig. 5-6). The benefits are also outlined in Fig. 5-6. More details on this type of planning and mapping, including costs, can be found in Dorney et al. (1985) and Dorney and Kitchen (1984).

A physiognomically (structurally) based biotic typology for all urban zones, with or without woody vegetation, has also been developed for planning purposes (Brady et al. 1979). The utility of this methodology is to

Table 5-2. Qualitative terrestrial vegetation evaluation system for urban areas (Dorney 1977).

Site region

Ecoclimate (normal, hot, cold)

Drainage D, DM, M, MW, W

Soil texture or organic 1 ts depth

Physiognomy: Pasture mowed, mowed lawn, rough, pasture, cropland by type, etc. Savannah by dominant species, woodland by dominant species (overstory, understory, etc.)

Average diameter, woodland overstory (6–10", 10–18", 18")

Age of overstory

Height of overstory

Percent crown closure

Production/respiration ratio

Resiliency (H, M, L) to drainage; to 50% crown opening; to compaction by man, to ground fire

---

Description and analysis of existing biotic resources

| 1 | 3 | 5 | 7 | 9 | 11 | 13 |
| 2 | 4 | 6 | 8 | 10 | 12 | 14 |

Capability

Not applicable in urban area

Functions provided by biotic resources of importance to man

| 1 | 3 | 5 | 7 | 9 |
| 2 | 4 | 6 | 8 | 10 |

Ability to control wind erosion H/M/L

Ability to control water erosion H/M/L

Contributes to bank stability (water, freeze/thaw)

Recharges groundwater

Protects discharge zone (springs)

Buffers/protects adjacent wooded unit from wind/sun

Microclimatic protection

Runoff coefficient (fractional number)

Energy buffer (noise, wind, water)

Filters air

Vegetation Analysis

History of vegetation management: logging, grazing, fire, maple sugaring, etc.

Wildlife use by order of birds, mammals

Acreage

Values/problems
of biotic resources

1 3 5 7 9 11
2 4 6 8 10 12

Overall aesthetic rating H/M/L

Uniqueness of biotic community in area H/M

Endangered/rare plant species list, or N/A

Endangered/rare animal species list, or N/A (vertebrates, all classes)

Buffers incompatible land uses

Provides open space linkage H/M/L

Potential for active recreation use H/M/L

Potential for passive recreation use H/M/L

Potential for nature education or research H/M/L

Production nuisance insects by type (e.g., mosquitoes H/M/L)

Damaged by pollution (explain)

IBP sites or equivalents

Table 5-3. Woodlot unit analysis showing various quantitative and qualitative attributes from an area near Toronto.

	Woodlot sector		
	15-E-(a)	15-E-(b)	15-E-(f)
Overstory composition[a] (species)	White Pine—30% Basswood—15% Ironwood—15% Sugar maple—10% White ash—10% Beech—5% Hawthorn—5% White elm Bur Oak Wild apple	Hawthorn—95% White pine White ash Trembling aspen	Red oak—50% White Ash—10% Shagbark Hickory—10% White oak—10% Ironwood—10% Sugar maple—5% Beech White pine
Stem diam. (cm)	15–30 30–45 (white pine)	5–8	30–45
Stand closure	70%	0%	85%
Understory composition (species)	Ironwood Sugar maple Hawthorn White ash Basswood	Hawthorn Raspberry Grasses Sugar maple	Sugar maple Ironwood Red oak Choke cherry
Relative age of stand	Immature to overmature	Mature	Immature

126

Drainage and soils	Moderately drained clays	Moderately drained clays	Moderately drained clay loams
Biological Health (H/M/L)	M	M	M
Ability to withstand (H/M/L):			
1) Wind stress	L	M	M
2) Construction traffic	L	M	M
3) Grade changes	L	M	M-L
4) Drainage changes	L	M	M
5) Intensive recreation	L	M	M
6) Passive Recreation	H	M	M
7) Edge manipulation	H-M	M	M
Sensitivity to urban development	H	L	M

[a] Overstory composition is derived from using point-centered quarter sampling technique. Three sectors are illustrated in Fig. 5-3.

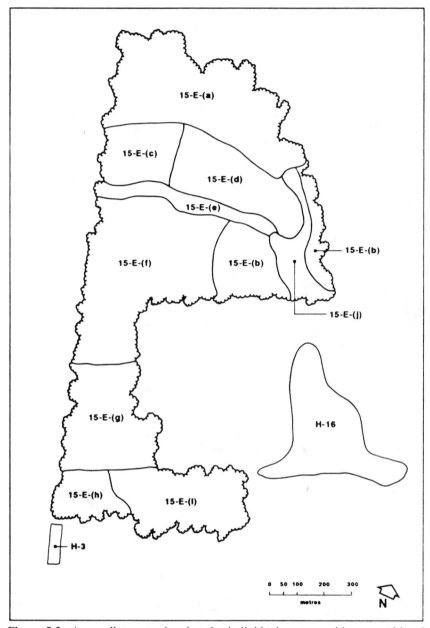

Figure 5-3. A woodlot map showing the individual sectors with compositional homogeneity. Note: Evidence of former pasturing by cattle in sector 15-E- (b) includes hawthorn, white pine, white ash, and trembling aspen. Courtesy of Ecoplans Ltd.

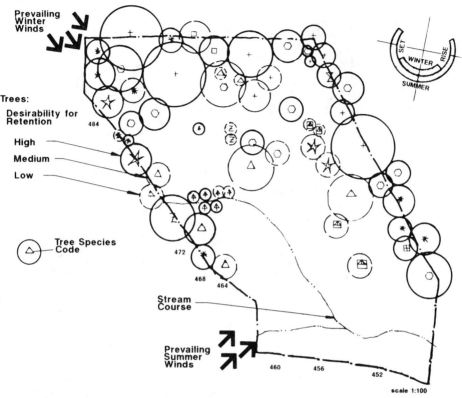

Figure 5-4. Site analysis for preservation of existing trees at site design scale. Computer tree inventory: Tree-by-tree inventory examines tree quality; biological health; suitability to site; and construction adaptability. Trees can be computer plotted at any scale. Site analysis: Site inventory examines soil conditions and suitability for development; stream courses, fisheries, and water quality; vegetation features (other than trees); and topographic analysis. Courtesy of Ecoplans Ltd.

relate maintenance costs, primary productivity, dereliction, rehabilitation, and wildlife populations to the built city and urban fringe, whereas the analytical techniques described above are focused on woodlot conservation and parks management (Dorney and Wagner-McLellan 1984; Dorney 1979).

Wildlife Survey Level of Detail

The relative importance of wildlife surveys varies according to use, whether for hunting or trapping, viewing, educational uses, and designation of sensitive areas. Usually such a survey is given low priority by most

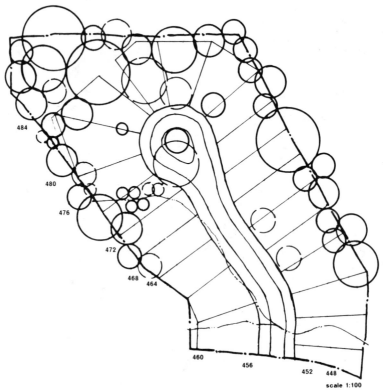

Figure 5-5. Road and lot design for minimizing loss of high-quality trees. Site design: lot and road layout based on topography and existing vegetation; trees not desirable for retention are removed prior to construction; lot grades and drainage modified to permit tree preservation; and, where possible, lot layout achieves energy conservation (i.e., summer shade, winter windbreaks). Courtesy of Ecoplans Ltd.

developers and government agencies dealing with urban and urban fringe problems. This is a sad legacy of traditional wildlife management with its emphasis on hunting and limited interest in nonconsumable wildlife. Some urban clients prefer to have no wildlife study done, because, in their opinion, it is an element that generates emotion from the public, often negative. In any case, there are some wildlife evaluation techniques that are applicable to the studies in the urban fringe and urban shadow zones. They are inexpensive and helpful at varying scales of detail where wildlife is likely to occur or known to exist. In wilderness, natural ecosystem areas, and agroecosystem areas, techniques for wildlife population assessment are found in Cowan's *Manual of Wildlife Techniques* (1958).

Generally, in and around urban areas, large game animals, such as deer, bear, wolves, and cougar, are extremely rare or do not exist and hence can be eliminated from analysis; deer may be an exception. It is the more

Figure 5-6. Final environmentally sensitive site design incorporating a site analysis (Fig. 5-4) and a road and lot design. The environmental site design benefits include community character (large trees and natural topography help define character of neighborhood), energy savings (proper tree location can produce annual energy savings of 15%, and proper house location allows passive solar utilization), marketing (large trees enhance the house site, increasing marketability), environment (trees filter airborne and dust contamination up to 75%, and trees reduce perceived noise levels from air traffic, roadways, schools), and cost savings (natural landscaping is cost effective, as you can save up to 66% over replanting costs). Note also that the value of the treed lots is about $5,000 more per lot than for non-treed lots. Courtesy of Ecoplans Ltd.

visible and uncommon species of birds—such as hawks, owls, plovers, eagles, herons, pheasants, ruffed grouse, quail—that need to be identified and whose potential to survive urban or industrial development must be evaluated. The smaller and less common species—songbirds, reptiles, and amphibians, which have primarily amenity and educational value— are not identified, because in most cases such information has limited utility in decision making. An exception would be if an endangered or uncommon species exists or is suspected to exist in the area.

If the regional-level faunistic level of diversity is done first, then local population density-diversity can be put into a conceptual context. Regional level statistics are available from bird atlas data in some countries (United Kingdom and Ontario, Canada) or singing bird transects (United States and Canada). For example, in the Ontario data, diversity by and large is not skewed by urban areas; the Great Lakes raise diversity more than escarpments do; some intensively farmed areas are low in diversity.

Localized breeding and feeding areas for large raptors or concentrations of nongame species, such as a heron rookery, a gull or cormorant colony, or the osprey nesting area that occurs in the village of Cedar Key, Florida, can be significant as "sensitive areas." As these areas are relatively small, any associated food resources and feeding areas must be identified for planning purposes, since they are icons to naturalists and become emotional issues if ignored.

In projects at an urban/regional scale (1:50,000 to 1:10,000), two approaches to wildlife evaluation provide a useful overview of wildlife populations. The first is to contact local naturalist clubs, naturalists, game wardens, and regional government biologists. The second is to examine specific winter and spring bird surveys, Christmas bird counts, bird atlas data, and spring breeding bird surveys. If more detailed population surveys are required, the client is apprised of the specific issue, its relative importance, and any time constraints due to seasonal patterns of breeding and migration. This way the client is more likely to fund the necessary field studies, which will be reasonable in cost.

For new town conceptual planning and subdivision conceptual design work (scale of 1:10,000 and 1:5:000), field surveys in winter and late spring conducted on foot and by car (using a window-mounted telescope along selected transects) or in particular habitats identify key wildlife areas, their probable relationship to present land use, and their potential for integration into new land use patterns.

For evaluation of landscape units at the new town conceptual design scale, I have found avian diversity indices to be too technical. Rather, for small areas I prefer to use food chain structure as a conceptual way of evaluating habitat quality for small areas. Food chains are constructed from breeding bird species lists; the technique avoids expensive quantitative population studies.

If the urban food chain is a detritus food chain, then it supports characteristic species such as ring-billed gulls, pigeons, or English sparrows, which may reach pest proportions. For an urban area of 250 acres or 100 hectares with 25 bird species representing consumers at the plant, insect, fish or minnow, and predatory levels, the avian assemblage is as close to one in an undisturbed marsh as can be found. The area is then mapped as a landscape type of significant natural or ecological value. On the other hand, if only red-winged blackbirds are found to be breeding in a marsh, since they derive much of their food outside the marsh, this may mean the water in the marsh is polluted or the marsh is too small or too simplistic in floristic structure to sustain a more diverse avian population. A simplified food chain structure or a detritus food chain structure signals that the habitat is damaged or of low quality. A complex food chain structure indicates that a more natural situation exists that can be given a higher value from a natural science perspective.

In one early subdivision design study (scale 1:5,000), we did quantitative breeding bird and mammalian snap-trapping studies (Kitchen 1969). In retrospect, by using vegetation or food chain structural analysis alone, we would have come to the same conclusion as to the various qualitative habitat types desirable for preservation without resorting to such costly quantitative wildlife studies.

For parks planning, for special educational site purposes, and for monitoring research (scale of 1:10,000 or 1:5,000), a detailed quantitative singing bird census may be desirable; quantitative mammal censuses and herpetological studies may also be required. Such work, because of cost considerations, is generally not done or is contracted over a period of years to hungry graduate students.

On occasion, detailed qualitative and semiquantitative wildlife studies can affect a design solution. A 1-square-mile university construction site at Utica, New York (scale 1:5,000), is broken up by two wooded river valleys. Winter and spring avian roadside censuses and walking transects not only reinforced my conclusion as to forest management needs but brought to my attention the desirability of maintaining some agricultural hayland on the proposed campus to preserve the nesting upland plover and sandpiper populations. This proposal also allowed one farmer to maintain a viable economic dairy operation, to return revenue to the state, and to provide visual corridors of high amenity value. However, it is infrequent, in my experience, that wildlife studies affect a design. This does not mean that wildlife studies should not be done but that before they are undertaken, the client should be aware that they may not achieve the payoff evident in geological, soils, and vegetation surveys.

Fisheries Surveys

Scale of effort for these ecosystem attributes is hard to define in the context of environmental analysis, since various techniques, such as netting and shocking streams, may be controlled by government regulations; they are costly as well. However, benthic invertebrate analysis (to family or genus level only) in streams has the advantage of being simple, not generally constrained by season or regulation, and repeatable monthly or bimonthly. The invertebrates, as indicators of water quality, temperature, and bottom conditions, provide a five-tier stream classification system ranging from "excellent" to "very poor" (Hilsenhoff 1977).

Changes in fish occurrence in streams and lakes over time can be determined through interviews of residents and by examination of archives. This information provides some overview and guidance as to possible landscape stresses for purposes of land use planning and management at the regional scale or on a watershed basis.

On two occasions, for the purposes of new town conceptual planning, fish shocking in streams identified brook trout habitat. However, this is expensive and requires careful control to avoid damage to the fishery. If shocking and fish habitat mapping is done, it can take 3 days for a two-person crew to cover a mile of stream sector by sector.

Water Quality Surveys

Water quality and stream and lake sampling require some knowledge of seasonal variability and an idea of what parameters are the most important to measure. A network of sampling stations is located. Usually two samples per month for 12 months gives an idea of seasonal variation and comparative variation across the network for sampling stations. For example, if potential eutrophication of an artificial lake is the issue, flow rate (cubic feet per second [cfs] or cubic meter per second [cms]), water chemistry, and chlorophyll analysis provide some insight into what quality of water a reservoir could contain depending on its size and depth. If the issue is impact of pipeline or highway construction on a trout stream, flow rate, sedimentation, and water temperature, not water chemistry per se, may be the more important parameters to measure. These are judgmental factors, of course. Because of the heavy cost in establishing such networks, parameters to be measured must be carefully planned and discussed with the client and approval agencies.

An aquatic ecosystem classification system for urban and urban fringe areas is a necessary planning and assessment tool. The one proposed in Table 5-4 describes the aquatic ecosystem using the same three-tiered system as in Table 5-2. The newer IBI (index biotic integrity) may offer promise as an alternative planner-evaluation tool (Steedman 1988) for both natural and stressed stream system.

Valley Land Surveys

For integrating urban valleys lands into a new town design at a scale of 1:5,000 to 1:10,000 or at the ecoelement scale, the ABC (Dorney 1977) has been adapted for the entire built and urbanizing fringe with emphasis on developing a user-oriented mapping system for both professional and nonprofessional audiences (Dorney 1978). It is currently used as a reference document for proposed development.

Table 5-4. Qualitative aquatic ecosystem evaluation system for urban areas (Dorney 1977).

Description and analysis of existing biotic resources

1 3 5 7 9 11 13
2 4 6 8 10 12

- Lake, stream, river, marsh, estuary
- Major fishery components, amphibian or shellfishery components
- Historical trophic status of lake
- Present trophic status of lake
- Present trophic status of lake
- Biochemical quality of river/stream (H/M/L)
- Spawning grounds, location, by species or family
- Migratory fish run by species
- Staging areas for waterfowl
- Breeding areas for aquatic birds; trophic levels present identified (I, I & II, I to III)
- Water level fluctuations in feet or meters
- Size of lake or marsh areas in acres, or sectional length/width of river
- Plant species
- Indicator invertebrates seen

Capability

Not applicable in urban area

Functions provided by biotic resources of importance to man

1 2 3

- Sewage assimilative capacity: degrades organic pollutants entering zone (biochemical sink)
- Traps sediment (marshes)
- Importance to terrestrial wildlife

Values/problems of biotic resources

1 3 5 7 9
2 4 6 8

- Overall aesthetic appeal H/M/L
- Major wildlife viewing resources YES/NO
- Major recreational gathering or fishing (food) resource YES/NO
- Hunting resource (ducks) YES/NO
- Provides outdoor education or research/scientific potential YES/NO
- Major swimming resource YES/NO
- Major boating resource YES/NO
- Biological nuisances—insect hatches, algal blooms, dead alewives, algal rotting PRESENT/ABSENT
- Hazardous area locations (engineered ditch), disease potential, bacterial and industrial pollution, location

We have also tried a simpler system of valley land classification to develop a high-, medium- or low-quality valleyland classification system for a secondary design level of detail. It is based on presence or absence of three factors: a minimum baseflow in summer or winter of at least 1 cfs/cms; presence of natural forest vegetation of 1 acre or 1/2 hectare or more in extent in the floodplain or valley walls; and a valley depth greater than 8 feet or 2.5 meters (deep enough to hide bikers and hikers). If all three factors are present, it is rated at 3, or a high-value area at an urban or subdivision scale (1:5,000); if only two factors are present, it is rated medium; and if one or none of the factors is present, it is rated low. This three-factor evaluation system provides a quick way to separate attractive wooded and enclosed valleys from engineered drains and swales. Bank erosion, outcrops of aesthetic or geological interest, nesting colonies of bank swallows, cultural or historical sites are additional features that could be mapped and included in such a simple ranking system.

These two valley land classification systems have satisfied hydrologists, engineers, designers, biologists, urban foresters, recreationists, and naturalists invited into consulting projects. Although these systems, including the IBI, are fairly new, they should encourage more work along these lines. Such urban valley land evaluation methodologies are not appropriate for mapping of higher-order rivers, such as wilderness wild rivers and major river systems such as the Niagara River and its Falls. Larger fluvial systems need evaluation at a national or ecoregion scale (probably 1:250,000).

Timing and Seasonal Aspects of Field Surveys

In the preceding sections some aspects of timing were alluded to. By way of amplification, the information in Table 5-5 detailing seasonal timing constraints for field studies clarifies this aspect. Many clients, unaware of natural rhythms, which can be partial or absolute constraints, expect that all information can be generated in a few weeks or in a few months at most, in step with other consultants' studies.

As a way to compress time, individuals like local naturalists, trappers, farmers, and historians provide insights and information on local historical ecology, land use, fauna, and flora, which, if verifiable, improves the quality of the analysis and provides good information at low cost. If the area is uninhabited, this assistance from local talent is not possible.

Phasing of the Work

If at all possible, environmental analysis should be broken into two phases: a reconnaissance level of analysis, followed by a more detailed analysis. If project complexity is high, this two-step approach is prefera-

Table 5-5. Seasonal constraints for ecological surveys.

Resource	Appropriate season for analysis	Constraints/remarks
Soils/geomorphology	Spring, summer (or wet seasons)	Frost, snow, or dry (hard ground makes field analysis difficult)
Geology	No constraints to drilling	Freeze-thaw action on banks needs identification in winter. Field work more difficult with snow on ground
Wildlife, migratory	Spring, fall	Ice; dangerous winds may hamper observation
Wildlife, breeding	Spring, summer	Highway noise can mask singing bird counts; rain and fog restrict visibility
Wildlife, wintering	Winter	Game counts use good tracking snow for observing food and cover patterns. Christmas bird counts are useful in North America. Pellet group counts for ungulates must be done before green-up
Fisheries, breeding	Spring, fall	Government regulations on netting or shocking
Fisheries, nonbreeding	Spring, summer, fall, winter	Government regulation on sampling. Critical temperature and intermittent effluent are important considerations; particular sampling days or times of day may be needed
Groundwater analysis	Spring and summer for peizometers installed by hand; drilling and pumping tests can be done at any time	Computer plots of well depths and water samples helpful in regional studies. Frozen soils may slow placement of peizometers
Climatology		Stable high-pressure area needed to measure inversion potential at mesoclimatic scale
Entomology	Spring, or at the height of nuisance season or breeding season	Wind, temperature, and diurnal activity patterns make study and sampling complicated

ble for reasons of cost effectiveness and budgeting control. For example, on a large hydroelectric project, a one- to three-person-month reconnaissance survey identifies clearly the issues to be analyzed and the methodology to be used. This phasing makes it possible to refine the budget and time to make it easier for high-quality work at the least cost. The reconnaissance phase usually includes an interim report as well as a detailed proposal for the second phase of the study. Opening the detailed phase to bids by other consultants tends to make the group doing the reconnaissance work "sharpen their pencil" when preparing the detailed work progress and budget. Hence this approach benefits the client.

The projects on which I have worked have not been of sufficient size or scale to warrant this phasing of work very often. However, the World Bank has used this reconnaissance approach at least on large projects. In view of the scale of investment, political-cultural sensitivity, and the complexity of analysis, it seems to be a prudent way to proceed. The quality and experience of personnel assigned to this phase should be the best available and, in theory, could act as a panel of peers during the detailed impact analysis and assist in the construction inspection and monitoring.

Social Science Surveys

The issues of scale, timing for surveys, and phasing of work discussed for various aspects of the natural environment apply equally to the social science surveys. Available statistical data provide an inexpensive and rapid first cut at many social and economic issues. This clarifies issues and allows for initial hypothesis formulation.

Then, as required, detailed questionnaires, community profiles, drop-in centers, surrogate groups, and interviews with respected community leaders are some of the additional devices to flesh out the initial sociological information base. Since costs for questionnaire data are high, the phased approach allows the client to be educated as the project evolves and the appropriate detail to be developed in concert with the client.

Issues of privacy, letters of introduction for the interviewees, and the need of reporting information back to the informants should be considered. In addition, some religious groups have strong patriarchal attitudes, making it nearly impossible for women to interview those men. These issues should be identified before interviews are undertaken, as they can lead to bad press, inefficiency, or distortions in results.

If public participation is an integral part of the project, this aspect needs to be carefully integrated into any social surveys and scientific field work. Once people are queried or see crews on site, a two-way flow of information is expected. This takes time, is expensive, and if poorly orchestrated, causes a political backlash.

Ensuring Confidentiality

When involved in government work where land purchase and land speculation are possibilities, confidentiality, security, access to files, maps, and reports need careful discussion with staff and tight control. The danger in unwarranted discussion or release of an unauthorized map can result in legal investigation and possible charges laid against individuals or against the firm. The reputation of a firm slowly built up over the years could be ruined instantly.

When government work is "touchy" from the point of view that it may affect policy and intra- or interagency politics, the confidentiality may be less rigid but nonetheless important. To avoid being blacklisted in securing future work, these "visible" or "invisible" lines should be probed before releasing any information to another agency, the public, and the press. Countries that have freedom-of-information legislation require that availability and publication of data be under the control of the contracting agency or corporation unless specifically spelled out by prior agreement. If publication is restricted, these rights should be clearly understood when a formal contract for services is drafted.

At the American Institute of Biological Sciences Symposium on Consulting Ecology in 1971, the inherent rights of the public to access to all confidential material, government or private, was raised by the participants; the public obligations and ethics of scientists were questioned in this regard. Needless to say there was no agreement; many felt no confidentiality was ever warranted. With this idealistic point of view, one can be sympathetic, but in my experience it is unworkable: it puts the environmental manager in an untenable position with the client. Especially for clients in the private sector, no continuing consulting work would be forthcoming if any consultants, be they engineers, architects, or planners, released documents to the public without authorization. The reason for this is clear. In a competitive market, the market intentions of a private company will often be blocked by competitors if the intentions are known before action is taken. Similarly, the documents may not represent the final position or the position the corporation will take in legal negotiations with approval agencies. I cannot conceive, therefore, how some professionals, generally academics, can expect environmental studies to be an exception to the usual situation where the client who pays the bill controls the information. There is nothing to prevent disclosure— as to when and how—if it is spelled out in the terms of the contract or required under a court order.

Perhaps a better way to handle issues of confidentiality would be by developing a public register of all environmental studies. Then guidelines regarding maximum time limits for confidentiality and penalties for nonconformance could result in an open system of reports. If clients failed to comply, they could be subpoenaed as long as the register was properly posted. Hearings to discover can identify such material as well, but by delaying report preparation or by splintering the report into memoranda, the reports can be effectively lost if the client decides it is in his interest to "lose" the report.

Ensuring Credibility

Maintaining credibility as a consultant is not as simple as it may appear to be; I have touched on this point throughout this book. Results of environmental studies are often not pleasing in part or in total to the client or prime consultant. When negative results are not expected or when they are potentially in conflict with the opinions of other professionals working on the same project, there may be peer and client pressure brought to change or ignore the results. This laundering of a report may ensure the short-term survival of consultants. Since credibility is based on performance over a much longer time frame, including the scrutiny of evidence to cross-examination in a hearing, to launder a report is self-defeating.

Some guidelines to assist in maintaining credibility when under fire are (1) double-check the analyses to be sure they are correct and accurate; (2) point out to the client that any alteration or deletion could be embarassing to all concerned, and if it resulted in property damage or loss of life, liability suits could become reality; and (3) suppression of data may be discovered in legal hearings, as testimony is usually given under oath.

Any technical errors should be corrected—if need be, at the expense of the firm doing the work. However, if the client insists on changes to alter or to cover certain findings, it would be preferable for the professional to ask to be released from any further obligation and to be paid for services rendered to date. If the matter were to be reviewed by a court or hearing board, the environmental firm would not have its credibility damaged. The difficulty with this latter course is that some contracts are tied to a final payment or hold-back (10 to 30 percent for acceptance of the final report). Although a firm's credibility is worth more than the 10 to 30 percent of a single contract, it may cause financial difficulties in the short term.

We have lost two private clients in 14 years over attempts at document alteration or because of disagreement with our final results. In the first instance, one consulting engineering firm under contract to a government agency hired a landscape architectural firm to redo some urban environmental impact work. In the second instance an as-built report was rejected because it was critical of a company's construction procedures, and our payments were withheld. By waiting for 8 months, a complete report was eventually filed over the Christmas holidays, while many of the company's staff were on holiday. In this case the company was up against an agency deadline for filing of January 1, so patience allowed the calendar to help ensure that an accurate report was filed. These situations are uncomfortable, but must be considered philosophically as part of the game. If the environmental advice arising from fieldwork is sound, it should lead to the same or similar conclusions if the same methodologies, scale of detail, and analytical processes are used.

Recently in Ontario, private studies were commissioned to check the adequacy of government reports and their ensuing regulations. One developer even contemplated commissioning a wildlife study in the hope it could be used as an argument to prevent landfilling of a marsh by a shopping mall competitor. Whether or not these motives and ethics of the private sector are condoned, it is not surprising to see environmental studies used for competitive purposes. Repetition of some studies may have the added bonus of improving the quality of work done by government, university, and private laboratories as well as eliminating inadequate professional scientists from the field. Using environmental studies to frustrate a competitor is unethical, and if that is the sole motive, it could lead to loss of a consultant's credibility.

Public action groups often attempt to use environmental studies to protect their amenity and property values by frustrating rezoning in an adjacent area. In a variation on this theme, wealthy people can use public action groups as a front to promote or protect their vested property interests. Again this is a questionable rationale for hiring an environmental consultant, and can affect the firm's credibility unless care is exercised by the consultant in understanding the clients' motives and representing the clients' interests.

In another vein, it goes without saying that meeting deadlines for reports is vital to credibility. In fairly complex projects where many subconsultants and special studies are commissioned, bringing it all together on time (and on budget) requires administration skills of a high order. Normally a flow chart or critical path analysis showing completion dates and interim and final reporting dates is prepared (see Fig. 4-2); any deviations must be cleared with the project manager.

Another issue of importance is the assigning of junior staff to undertake work when senior staff were indicated as being involved directly at the time the work was commissioned. Clients and associate consultants become suspicious if this practice is followed too frequently. Some agencies like the World Bank require a signed letter by the principal that those staff members indicated on a proposal will actually be available and will be undertaking the work; presumably this requirement arose because of problems resulting from staff substitution.

Credibility of the environmental consultant is ultimately tested in the board room of the corporation, in the minister's or his deputy's office, in the courtroom at the hearing, or under the eyes of the press. The old adage, articulated by Harry Truman, "If you can't stand the heat, get out of the kitchen," applies to the issue of credibility: it is not easily attained, but can be easily lost.

Legal Liability

As this section is not written by a lawyer, an initial qualification is necessary. On specific issues, the question of liability must be answered by a lawyer. What follows is based on experience and is intended to emphasize the importance of this aspect of professional practice.

Legal liability generally falls into two categories—errors of commission and errors of omission. The former are erroneous conclusions or fallacious results, while the latter are those issues that were not perceived as important or not considered. Two examples demonstrate the difficulties that can arise from both situations.

In the first, briefly described earlier, the environmental assessment for a Canadian pipeline project included a description of terrain or geomorphology. Since pipes are to be buried at 5 feet or $1\frac{1}{2}$ meters, any bedrock closer than this to the surface requires blasting or "ripping" — an expensive procedure compared to the cost of trenching in soil free of stone. As the environmental report indicated no bedrock constraints, the contractor bid the job at a lower cost. During construction, bedrock was found, so the contractor asked for additional monies or "extras." The pipeline company then went to recover these "extras" from the environmental consultants, as their report had misled the contractor. The amounts of money involved were substantial. As professional liability insurance for environmental consultants is not available in some jurisdictions at this time, a large engineering firm may find its overall resources drained, and a small environmental firm may be forced out of business.

The second example was a housing project in Virginia in the 1970s. A large engineering firm bought a small environmental consulting firm and inherited, along with its clients, a report whose soils map was incorrect. The conclusion drawn from this map was that the soils had a low potential for erosion. Since potable water was an issue, erosion could impair the drinking water quality. This mapping and interpretation error resulted in a six-figure law suit for damages due to delays in construction and the need to redesign the site to reduce the erosion. Expert testimony of a soil scientist hired by the county was all that was needed to stop development and require a new study. The damages were settled out of court (J. Hackett, personal communication).

As North American clients in this decade seem more willing to take a variety of professionals to court, environmental consultants will not be overlooked. Accuracy, attending to detail, asking the proper questions, and being able to defend the work before hearing bodies assist in preventing liability claims. An additional complexity is introduced, however, when many separate consulting companies work on a project: all may be swept into court, with the court determining the proportionate liability. In these cases it may be expensive to demonstrate to the court that no liability is involved.

In doing environmental surveys and environmental inspection, liability insurance with regard to property damage is useful. Liability cannot be as dramatic and financially crippling as in the above two examples: a soil scientist may leave a farm gate open, and the escaping cows may be hit by cars. Coverage for this type of eventuality is inexpensive. Some companies routinely require environmental consultants to carry such liability insurance, up to $1 million, before undertaking the work.

Preparing for Public and Judicial Hearings

As the issues of environmental quality have become politicized and popularized by TV and the press, scientists and environmental managers and other expert witnesses (in North America) have become increasingly involved in litigation and public inquiries. In Canada, the Ontario Municipal Board (OMB) and the Environmental Assessment (Hearing) Board, Land Compensation Board, National Energy Board, Ontario Energy Board, and Consolidated Hearing Board, among others, as well as the courts, are some of the bodies in which planning issues and projects of major public works are scrutinized in a legal forum. There are counterparts in Britain and the United States. Informal or formal cross-examination is legitimately part of the procedure. Environmental management concerns then become one of the foci in the arguments.

We have found, as have other environmental consultants, our work increasingly subjected to scrutiny by such judicial bodies. In the case of the OMB, some of our early work for a new town in Ontario used a qualitative rating of woodlots to rank their potential for parkland. In 1973 it was subjected to scrutiny by rate payers who wanted more of the woodland dedicated as parkland than the town had agreed to purchase. To our surprise, we found the rate payers had commissioned a separate quantitative ecological study. Fortunately, from our point of view, the two studies, although differing in level of detail, came to the same conclusion about portions of the woodland that were of highest potential value for parkland. When the conclusions are dissimilar, the painful dissection of methodology, analysis, and eventually credibility of the analyst are examined by lawyers who are well trained for just this situation. Errors of omission or commission will be uncovered.

At the outset of a contract, it is important to tell the client that the analysis and a methodology for generating the data will be credible before a court or in a hearing—that they are procedures designed to answer any public criticism of both a general or specific nature. The client's legitimate interests will be protected, as will, it is hoped, the firm's credibility. The open public hearing and formal judicial hearing have the effect of making all environmental consultants much more careful about doing "quick and dirty" analyses. Such hearings also have made it easier to convince potential clients about the facts of life without antagonizing them or making them think the consultant is asking for unreasonable fees and long lead times to do appropriate professional work. Since hearings on an urban development proposal, for example, may take place 3 to 5 years after the field studies have been completed, it is vital to revisit the site, to check if the biophysical inventories are still correct, to understand any changes in governmental policy or plans, and to identify any design or servicing changes undertaken by the private sector that could affect the original environmental report—before appearing at the hearing.

A few weeks prior to a hearing, the witnesses normally meet with their lawyer. At this time the witnesses' statements, if any, are reviewed, and any interrogatories (formal questions asked of the opposition to clarify issues before the hearing starts) are drafted; the project documents are reviewed, and all options and alternatives are reexamined. Since the client places its case in the hands of the lawyer, a fresh, somewhat independent point of view develops as the lawyer tests the arguments. At this time, more expensive but better environmental options that may have been rejected can be reinstated as bargaining chits or simply conceded at the outset of the hearing. The reason that this last-minute flexibility may develop is simple: lawyers do not like to lose. They can often be persuasive where an environmental manager has not been persuasive. Also they need to be aware of any tradeoffs, since bargaining may be beneficial.

Privileged information generally means that information prepared specifically for a hearing is to be kept confidential between the lawyer and the client. For example, evidence in regard to possible settlements may be privileged. What evidence may or may not be privileged is an important matter requiring clarification within the team before the hearing commences.

The intervenors (objectors) in a hearing often have specific concerns. If these are matters relating to environmental quality, rather than to economics, the environmental manager is expected to anticipate the depth and breadth of the attack and to be prepared to give his or her lawyer questions for the cross-examination of the intervenor's witness. Often I research the background of an intervenor, particularly if it is a well-published consultant or academic who represents the intervenor. Past publications offer clues to how the individual conceptualizes a problem, the methodology likely to be used, and the general or specific conclusions that can be predicted. A botanist sees the world in meter squares; an ornithologist sees a smaller-scale world by looking upward into the tree or across the meadow.

Letters of comment, petitions, or written submissions may be allowed from any affected party. Any such communications would be available to all parties. Documents related to witnesses' testimony are made available at a reasonable time to allow all parties to review them. One to 2 months is usual. If less than this time is available, a hearing may need to be postponed.

Negotiation or third-party mediation prior to a hearing can be cost-effective. Its cost-effectiveness results in reducing the length of the hearing or in eliminating some intervenors completely by conceding to their real or perceived desires. Presubmission consultation where all interested parties discuss a project informally at its inception and agree to agree or to disagree can also reduce the eventual length of a hearing by removing some uncontested evidence (Dorney and Smith 1985). If the parties sitting down at the table are not the total number, those not present have certain "natural rights" that other parties cannot unilaterally abrogate. Summons can be issued to require knowledgeable people to appear and testify.

For those unfamiliar with legal hearings, they generally follow this sequence of steps:

1. In a large complex hearing, an exchange of interrogatories may precede preparation of the witness statement (summaries of evidence to be led). The purpose of this exchange is to focus on technical gaps in evidence reducing the potential for trial by ambush. Each side may try to flush out a set of calculations or assumptions in an analysis, so that recalculations, if needed, can be properly prepared. It is a kind of sparring match before real blows are landed.
2. Prehearings may be called to define issues, including interrogatories and witnesses' statements and procedures for conduct of the hearing; issues raised at the preliminary hearing may be settled by an interlocutory court order or by minutes of settlement.
3. Next is serving of notice to affected parties. Some regulations or laws require notification of owners or persons within certain distances of a site; for this purpose registered letters and newspaper announcements are used or required.
4. Filing of documents, if required, has time and place limitations imposed.
5. Notice of motions may be heard prior to the opening of the hearing; such motions ask for "relief" from certain conditions—for example, adjournment to allow more time to study documents.
6. Procedures are established for preparation and release of transcripts (if any).
7. The opening of the trial or hearing begins by introduction of all lawyers to the hearing board or legal panel (name, firm, address).
8. Identification of intervenors to legal panel is then done for those not represented by legal counsel.
9. All parties agree as to sequence of evidence to allow witnesses to be scheduled efficiently (to help to reduce costs).
10. The hearing board may visit the site.

11. Presentation of case then takes place by the project proponent:
    * History of project (relevant statutes or relevant policy (by lawyers).
    * Presentation by lawyers of vitaes of witnesses to qualify them as "experts."
    * Swearing and qualification of lay and technical witnesses or panel of witnesses.
    * Evidence by technical expert or experts, separately or in a panel (evidence in chief); maps and reports given exhibit numbers.
    * Cross-examination of evidence in chief as witnesses appear.
    * Reexamination of witnesses if lawyer desires to clarify any points raised in cross-examination.
    * Cross-examination of any points raised in reexamination, if desired by opposing lawyers.
12. Presentation of case then takes place by intervenors:
    * Swearing and qualifications of witnesses for intervenors.
    * Evidence by experts for intervenors (evidence in chief).
    * Cross-examination of evidence of intervenors by proponent's lawyer.
    * Reexamination of witnesses if lawyer desires to clarify any points raised in cross-examination.
13. Reply evidence or rehearing of new evidence from any parties may be allowed, but must first be requested and a rationale established for this request.
14. Final arguments or summation by lawyers, based only on evidence presented, then completes the hearing.
15. Completion of hearing.
16. Hearing board's decisions/findings are then released, generally a few months later. For simple cases a ruling may be made right at the conclusion of the hearing.

To be a good witness is simply to be a good communicator:

* Speak slowly and distinctly so the legal panel and the court transcriber can follow your evidence; place names or people's names can be troublesome, so spell them out as required.
* Answer questions thoughtfully and preferably with short answers that are unambiguous, such as "yes," "no."
* Face the judges, so they can hear your evidence.
* When judges are writing down some of your points, it means your evidence is meaningful; when this happens, slow down, so the judge can transcribe his comments fully before moving on to new evidence.
* Remember all dates when work was performed and all people contacted by phone or in person.
* Familiarize yourself with all geographic places and names in the study area.
* Do not establish eye contact with the intervenor's lawyer. Keep eye

contact with the judges, forcing the intervenor's lawyer to "pace." This gives the witness an advantage of appearing confident; intervenors are thrown off balance and appear frenetic.

- Review all previous work on similar projects, and be prepared to defend any differences.
- Keep body relaxed; do not show nervous tension by fidgeting with pencils, thumbing reports, or jingling coins in your pocket.
- Refer only to documents or reports that you are prepared to have become property of the hearing. Intervenors' lawyers are quick to ask to see any documents a witness refers to, as it may help their case or make it appear that the proponent has suppressed information.
- Questions beyond your level of expertise should not be answered, as it will damage your credibility as an expert witness, or be answered only after stating that it is beyond your specific knowledge or that another witness will deal with the matter later.
- Questions of depth deserve a thoughtful answer, so to stall for time, ask the lawyer to repeat the question or to clarify the question. This delay confers an advantage to the witness, who otherwise may become entangled in a complex answer and diminishing accuracy and credibility.
- Refer to journal articles and invited papers given at seminars to show you are a witness of substance; such a tactic builds your credibility, putting the intervenor at a disadvantage.
- Discussion with your lawyer when recessed during cross-examination is improper, so drink your coffee or eat your lunch alone.
- Read your testimony as transcribed (usually the next day). If you are depending on gestures, jargon, or the use of partial sentences to convey meaning, this becomes self-evident and can be improved by rehearsal for subsequent hearings.
- Find out the technical expertise of the judges. If one judge is a key individual, orient your evidence toward that individual.
- Discussion with the public, press, or the intervenors' lawyers during recess while on the stand is improper and imprudent.

Good witnesses are people who can be both precise and concise. Professors or teachers are generally poor witnesses, as they continually qualify statements, and they do usually not know when to stop talking.

Establishing Fees

Admittedly, this is a difficult issue to define with clarity. However, the following fee schedule is similar to the one used by many consultants in Ontario and the United States. A markup on employees' salary of 2.4

times covers pension, unemployment and insurance benefits, vacation time, sick leave time, and general office overhead; it should return a 10 percent profit to the firm on gross receipts. Table 5-6 contains rates for various levels of academic training and experience.

A professional practice generates overhead; about $15,000 a year would be the minimum for a small office of 500 to 700 square feet and with a part-time secretary. An office of 1,000 square feet (able to house a staff of four or five persons) with the salary of a full-time secretary makes overhead close to $40,000 per year. A certain amount of "down time," when the professional staff are developing contacts with prospective clients, are on vacation, are attending meetings, and have no contract work is expected. Down time should be kept under 20 percent, including vacation time. For these reasons, daily or hourly rates may seem high, but professional employees cannot always be working on contracts, and overhead never stops. Sick leave, vacations, employee benefits, severance, and expenses to attend professional meetings must also be met out of gross revenue.

When one is asking top rates in the field, $500 to $700 per day or more, clients may expect much of the fieldwork to be done by less experienced help. Furthermore, they may suggest that monthly overhead be held so as not to exceed a maximum amount, based on the percentage of total work being done for them, as offset against total monthly overhead. In some countries feather-bedding or kickbacks are usual; since these are usually illegal in North America, great care should be exercised before getting into this situation. Advice from lawyers and tax experts is critical.

In my experience, experience with technicians, with students at the third-year level of academic training, and with those having a bachelor's degree and no experience suggests that such personnel cannot be technically proficient to qualify as professionals without five or more years of responsible technical and supervisory experience. Hourly rates charged clients for such full-time technicians are marked up by 2.4 times for full-time personnel and 2.0 for those on part-time salary. The lower

Table 5-6. Rates to individual client.

	Salary	Markup
Bachelor's degree, no experience	$ 7/hr	$17/hr[a]
Master's degree, no experience	$10/hr	$25/hr
Master's degree, 2 years' experience	$15/hr	$36/hr
Ph.D. degree, no experience	$15/hr	$36/hr
Ph.D. degree, 2 years' experience	$25/hr	$60/hr
Master's or Ph.D. degree, 10 years' experience	$30/hr	$72/hr

[a] These costs reflect hourly rates based on 1984 prices in Canadian dollars ($1.00 US = $1.32 Canadian).

markup on such part-time people results from the situation that fewer benefits are required to be paid to them.

When projects tend to run over budget, the usual way to keep them in the black is for the senior personnel to "donate" their time at a proportion of what they usually receive. Clients are sometimes sympathetic to overruns and sometimes not, but I have found it pays to submit an actual bill to them for the time spent and expenses, showing an appropriate discount, so you have a record when estimating future jobs. Furthermore, when done this way, it demonstrates to the client that you did not cut any corners on quality to stay within a previously negotiated upset limit. If you are a subconsultant, the prime consultant may have unexpended funds that come from a contingency fund or which accrue from a shortfall in other parts of the budget. It pays to notify the prime consultant as soon as a budget overrun is anticipated and tactfully ask if supplementary funds are available.

Determining Cash Flow

Finally, operating an environmental practice requires not only scientific expertise but management skills for determining cash flows and financial operations. For those not well acquainted with operating a professional practice, management of cash flow can be not only a continual headache but a disaster if not controlled. If the firm is launched on a shoestring, accounts can have an agonizingly slow turnaround. Specialists subcontracting to do specific scientific field or library work generally ask to be paid weekly, monthly, or on acceptance of their section of the report. By the time the environmental firm completes the final report, usually another 2 to 4 weeks, and then submits it to the prime consultant or directly to the client, further delays are inevitable. For this reason it is useful to bill the client monthly, expecting a turnaround in 60 days. Thereafter interest of 1 to 2 percent per month is a legitimate charge.

Because of these financial realities, a line of credit for $20,000 to $50,000 (or more) from a local bank for a small to medium-size firm is useful. A financial agreement with a better-endowed firm can help cover deficits in operating capital, and start-up costs on large projects. An advance on a large project for start-up costs is also reasonable. In addition, to minimize the need for credit, principals in a firm can agree to be paid when payment is received, not when billed.

If annual cash flow is in the order of $200,000, credit up to $20,000 to $50,000 should be sufficient, assuming principals are only paid when money is received. For larger contracts in the order of $50,000 and with annual cash flows of about $300,000 to $750,000, credit to $100,000 to $200,000 is necessary. If operating capital is held in the firm, this reduces the need for a line of credit but may attract different tax rates, since it is taxed at the usual rate of corporate profit, and interest on money borrowed may be deductible as a business expense.

To develop operating capital, it is useful if the principals can leave their fees in the firm as gross profit. However, to remove the money later may mean that it is taxed twice. Otherwise, to sustain growth, the firm may be recapitalized to get some financial leverage or flexibility. Its treasury shares may be sold to silent partners to maintain stability and to achieve a more rapid rate of growth if desired.

If a banker withdraws the firm's line of credit, insolvency may result. The reasons for doing so may be that accounts receivable over 60 days are too large, or it may be due to a loss of faith in the officers or due to personal financial problems of an officer, which may affect the solvency of the firm. If a line of credit is used, continual dialogue with the bank manager and monthly statements are essential ingredients to develop mutual trust. Signing personal guarantees to pay all loans can be expected of the principals. This means, of course, that all of a principal's assets, including car, house, furniture, recreational property, securities, and cash, can be seized. To avoid this need for guarantees, it helps to have value left in the firm (equipment, cash, receivables, securities) to cover at least part of the line of credit. It is important to realize that bankers have some flexibility, but limits are imposed by their directors and by regulatory bodies.

Bibliography

Brady RF et al. (1979) A typology for the urban ecosystem and its relationship to larger biogeographical landscape units. Urban Ecol 4:11–28.

Cowan IMT et al. (eds) (1958) Manual of Wildlife Techniques. Washington: Wildlife Society.

Dorney JR, Leitner LA (1985) Woodlot scale: A method for rapid assessment of woodlot values. Environ Mgmt 9(1):27–34.

Dorney RS (1973) Role of ecologists as consultants in urban planning and design. Hum Ecol 1 (3):183–200.

Dorney RS (1977) Biophysical and cultural-historic land classification and mapping for Canadian urban and urbanizing land. Proc Workshop on Ecological Land Classification in Urban Areas, Can Committee on Ecol Land Class. In: Land Class Series No. 3. Ottawa: Environment Canada, pp 57–71.

Dorney RS (1978) Urban Valleylands Study of the City of Waterloo. Waterloo, Ont.: University of Waterloo, School of Urban and Regional Planning.

Dorney RS (1979) The ecology and management of disturbed urban land. Land Arch (May):268–272,320.

Dorney RS, Evered B, Kitchen CM (1986) Effects of tree conservation in the urbanizing fringe of southern Ontario. Urban Ecol 9:289–309.

Dorney RS, Kitchen CM (1984) A Tree Saving Manual for Developers, Builders, Designers, Arborists, and Landscape Contractors. Working paper No. 18. Waterloo, Ont.: University of Waterloo, School of Urban and Regional Planning.

Dorney RS, Smith LF (eds) (1985) Environmental Mediation. Working paper No. 19. Waterloo, Ont.: University of Waterloo, School of Urban and Regional Planning.

Dorney RS, Wagner-McLellan P (1984) The urban ecosystem: Its spatial structure, its scale relationships, and its subsystem attributes. Environments 16(1):9–20.

Hilsenhoff WL (1977) Use of Arthropods to Evaluate Water Quality of Streams. Technical Bulletin No. 100. Madison: Wisconsin Department of Natural Resources.

Kitchen CM (1969) An ecological approach to improving urban design. Unpublished MA thesis, University of Waterloo (Ontario), School of Urban and Regional Planning.

Lang R, Armour A (1980) Environmental Planning Resourcebook. Ottawa: Lands Directorate, Environment Canada.

Steedman R (1988) Modification and assessment of an index of biotic integrity to quantify stream quality in southern Ontario. Can J Fish Aquatic Sci 45 (1):492–501.

6
Report Preparation and Presentation

With almost every environmental management consulting study under-taken, the client expects a written statement of findings and recommenda-tions. Hence, this chapter delves into this important aspect of profes-sional practice. A report of this kind differs in some particular details from the general field of technical writing because of its interdisciplinary nature and its need to communicate to a varied audience in a socially effective and cost-effective manner.

Irrespective of the length, the report serves as a permanent bridge or link of communications between consultants, the client, lawyers, and often the general public. The report's importance cannot be overem-phasized. In the short term, the report justifies the time, effort, and cost allotted to the environmental consultant and subconsultants. In the long term, the report serves as a benchmark of the quality of work done by the consultant or the firm and as a working document if and when the project proceeds. Over the long term, the eventual distribution of the report can never be anticipated. A poorly done report can, through circulation to other professionals and the public, be an unpleasant reminder, difficult to live down. On the other hand, a well-done report serves as subtle advertising to those who continually or occasionally refer to it.

Unfortunately, most science and engineering professionals have no formal academic training in the field of communications—that is, in interesting and tastefully done graphics and maps and in jargon-free English, for verbal and nonverbal presentations. It is a rare science or engineering department that offers graduate courses in report writing, design, or graphics; students absorb these skills by osmosis if at all. The absence of such courses signals to graduate students that communication skills are secondary or of no consequence at all. This contrasts sharply with training of planners, landscape architects, and architects, who devote from 10 to 60 percent of their educational time to development of graphic and verbal communication skills, and to social scientists, who through debate and participation in public meetings develop, in most cases, verbal skills.

Communication then can be an obstacle to be overcome by some technically trained professionals, particularly if a quality practice in environmental management is to be established quickly. It is not unusual to find a client who has had poor experience with technically trained professionals, irrespective of their technical competence, simply because their reports are hard to understand: too technical and loaded with jargon or with confusing graphics. Hence the quality of the report becomes an irritant to other professional readers and negates the inherent value the report would have had if attention had been paid to these aspects.

Not infrequently, a client who understands the time, effort, and amount of money required to produce a high-quality technical report asks the environmental management consultant to write only a short summary or a list of recommendations. Since the fieldwork and data synthesis are to most natural scientists the more enjoyable part of the effort, it is tempting indeed to accept this proposition. However, as in most situations, there are two sides: the pitfalls must be weighed against the advantages.

The *pitfalls* in not writing a reasonably complete report, but in giving an oral presentation with only maps and tables are these:

1. Many years after a job is completed, clients frequently ask for additional evaluation or additional interpretation based on new options for development or facility use. The report, if it is complete, acts as a quick and accurate reminder of past and present ecosystem structure and processes observed, site construction conditions during development, the methodology used, any value judgments made, and any verbal or written recommendations made. If material is only in the form of rough field notes and assorted field maps, a fair bit of time is required to reorganize the material into a report the accuracy of which, owing to the intervening years, may be impaired.
2. The personnel who did the fieldwork and/or maps and tables may no longer be working for the firm. The supervisor, if asked at a later date to comment on the work, may guess what was done, what was said, and what the data mean.
3. If a public hearing or meeting is called, lead time to prepare for the hearing may not be sufficient to produce a final document, including maps and high-quality graphics. Maps, slides, and overhead transparencies can take up to 2 to 3 weeks to prepare and be copied.
4. Parts of a report, if completed and reasonably clean, such as figures,

tables, or drawings, can often be updated with limited alteration. A complete initial report facilitates supplementary work and subsequently reduces the time and cost to the client.

5. Misinterpretation of findings by a client or by his other consultants is less likely if a complete, well-edited report has been prepared. A person's memory of what exactly was discussed and agreed upon months or years later is characteristically hazy. In addition, some of the subconsultants and staff may no longer be readily available for assistance, requiring hasty data reinterpretation by other staff. If legal liability is involved, a report—or lack of it—can become crucial evidence affecting the environmental manager's credibility.

The master copy of all reports should be filed separately, complete with maps. Desk copies become marked and torn; their maps can disappear. If desk copies are borrowed by other staff, they eventually end up in strange places—never available when needed. In addition, clerical staff can quickly retrieve the master copy if it is filed in one place, never circulated or removed from the office.

At the time a project and budget are under negotiation, it is crucial that the environmental manager thoroughly explore with the client the nature of the report expected. Any additional time and money required for a comprehensive and well-illustrated report must be assessed accurately, including the likely target audience(s). Unless the nature of the report is clearly articulated, deadlines and budgets are easily exceeded, jeopardizing quality and effectiveness of the entire work effort.

Briefly, this is the nature of the problem and the reason for devoting an entire chapter to report preparation and presentation. Not only are technical and organizational skills required to operate an interdisciplinary consulting practice, but communication skills are required as well. Since "beauty is in the eye of the beholder," the images projected by the environmental firm through their reports are crucial to effect a successful courtship and a successful consummation.

Expanded Role of the Communications Specialist

As presented earlier in this book, I see the ecosystem analysis as having to reach deeply into the social, biological, earth, and engineering sciences to achieve any fundamental understanding of ecological processes. However, a potential nightmare for effective communication is created, since each separate compartment of the sciences has its own specialized vocabulary. Linguistic scholars recognize that group identity is reinforced by developing a specialized vocabulary—"secret rites" into which a specialist is initiated in graduate school. Although such specialized vocabulary serves the purposes of a group, it clearly runs counter to effective intergroup discussion, agreement, and action. A communications specialist's role is to assist the environmental manager by acting as a translator, that is, "laundering" the jargon without muddying the meaning. As any translator can verify, it is not an easy task to translate technical material accurately that has sociopolitical and economic context.

As well as acting as a technical translator, the role of the communications specialist ideally extends to the more conventional area of report structure, page layout, and graphics supervision. With little additional effort, a report can be presented attractively. Illustrations and pictures break the monotony of pages of written text, improve retention of concepts, and project a positive message. An attractive cover invites a reader to pick up the report and read it.

Incomplete and inaccurate citations of references also jeopardize the credibility of a team effort. The communications specialist has to ensure that quotations are accurate, footnotes and bibliographies are complete, and all citations in the text do, in fact, occur in the bibliography. Since it is common to "borrow" information from other in-house reports, proper documentation helps verify the source of technical information prior to a legal hearing. Plagiarism or perceived plagiarism can be deadly to professional credibility if large sections are borrowed from other reports without acknowledgment.

Viewed in this fairly comprehensive way, the communications specialist facilitates document preparation, fills the the more or less classical editorial role, and translates the specialized language of the sciences without distorting their meaning. The translation role, then, elevates this specialist to a crucial position in the professional team. Ideally the communications specialist should be independent of but integrated operationally into the interdisciplinary team.

To achieve integration, the communications specialist should be brought in, as noted earlier, when the project is first beginning. By consistently meeting with the subconsultants at the time the groups are discussing the relationship between environmental resources and the issues at hand, the communications specialist develops a feel for the subject, the specialized concepts, and the specialized language associated with these concepts. When the subconsultants submit their findings, the editor/translator reorganizes their sectional reports, removes or footnotes any technical terminology, and relegates repetitious material to an appendix.

If the communications specialist is brought in only at the final stages of a project—this is done principally to save money—the responsibility for document preparation falls on the shoulders of the project leader, generally the senior environmental manager. The work of the communications specialist is reduced to the usual role of doing page layout, bibliographic verification, and proofreading. This makes it somewhat harder to produce both an accurate and a readable report, since a single technical professional tends not to perceive his or her own linguistic limitations; an independent communications specialist provides the degree of independent judgment needed.

The kind of person who can take on this communications specialist/ editorial role is hard to define in precise academic terms, but it is generally someone having language or English training combined with some artistic skills and grounding in a science. In addition, he or she should be able to work well with technical professionals and under the pressure of deadlines. As deadlines grow near for completion of a project, unless the communications specialist has stamina and a low "frustration index," the first job may prove to be the last.

Timing

A reasonably well illustrated final document needs sufficient time and a few extra days as insurance against unforeseen problems. For example, a report 150 to 200 pages in length usually has three to five maps and typically five to 10 figures or charts. It may contain six chapters of which four are written by subconsultants and the remainder written in house. Assuming a project that extends over 8 full months, by the beginning or middle of the seventh month (6 to 8 weeks before the deadline), the subconsultants should be completing their fieldwork and ready to submit maps, drawings, and figures. If they cannot meet this first deadline for graphics, the project manager states the desired scale, color, legend, and layout clearly to them so that when their tardy work is forthcoming, it does not require last-minute redrafting. This avoids overloading the draftsperson, who may have to go into overtime work, allow the graphic quality to suffer, or simply not meet the deadline. Figure 4-2 illustrates a more complex 18-month project requiring three separate reports, all of which require coordination as regards technical information and graphics.

Also near the middle or at the end of the seventh month, on the same hypothetical 8-month project, the project coordinator, with the help of the communications specialist, prepares the report outline. An outline includes details on the sequence of chapters, chapter length, relevant maps and figures, appendices, and any special graphics for the cover.

By the beginning of the eighth or final month, the first drafts of the separate chapters come from the subconsultants. The communications specialist then restructures the language, as needed, checking with the subconsultant and project leader regarding any revisions, deletions, and additions. Final typing and proofreading, chapter by chapter, proceed smoothly throughout the first 2 weeks of the final month.

By the third week the assembling of the chapters is done, including the tables, graphics, and maps in their proper places. The summary and recommendations are prepared jointly by the communications specialist and the project manager as the last step. Also, in the final week, or in the final days, numbering of illustrations and pages is done. A final proofreading is completed, and duplication proceeds. One day for page collation and binding sees the manuscript into the mail on time. For a manuscript of the size specified earlier (150 to 200 pages), I allot approximately 1 month of time for the communications specialist, about 80 percent of it during the final month of the project. (The additional cost for including such a communications specialist is discussed later in this chapter.)

If the subconsultants fail to meet their deadlines and many of the graphics are not prepared before the final month, the entire manuscript effort tends to pile up 2 weeks before the deadline. Such a log jam requires heroic efforts to avoid a 7- to 10-day overrun of the deadline. An overrun disrupts other projects and infuriates clients, especially if submissions to approving committees have hard deadlines or hearings are about to begin.

When a large number of copies is required, normally a printer does them. Since availability of printers varies from place to place, the amount of lead time necessary to make a large run varies from a few days to a few months: color runs require longer lead times. If such large runs and specialized printing are needed, the additional time must be allotted, with the larger commercial run to be delivered later. However, in-house printing and word processing technology have improved considerably in recent years, making available numerous methods of report production.

If careful editing of a report is not feasible, or if a client wants to review a report before it is finalized, stamping "DRAFT" on the report cover, recommendations, and executive summary is usual procedure. Technically, this meets a deadline and indicates that a second approved version is to be expected. Cleanup proceeds at a more leisurely pace.

Text Organization

Although every office sooner or later settles down to a distinctive pattern of report layout and preparation, it may be helpful to point out what I think are key elements in a consulting report to create a professional appearance.

The cover of a widely circulated report should be designed by someone with an artistic flair. A photo or line drawing is integrated with the title and the consultant's name. Color is preferable but usually requires more lead time, possibly 2 to 3 weeks for a press run of 25 to 100 covers. The cover design and its printing can be done early on in a project.

The letter of transmittal is the first page of a report. It states for whom the report was prepared and the general terms of reference. It is dated and signed by the project manager. Then comes the table of contents, followed by an executive summary, which highlights the key parts of the work and which cross-references them to the appropriate text page. This is followed by the recommendations, which are usually numbered and outline the action program chosen. For easy access, each recommendation also should be cross-referenced to appropriate tables, figures, and text material.

The rationale for putting the executive summary and recommendations first is that many of the readers—other consultants, lawyers, clients, administrative staff—are busy people. They usually take time to read the summary and recommendations and pursue those specific recommendations that either surprise them, affect them in some way, or run counter to their current understanding of the problem. Cross-referencing allows scanning of specific material and studying it as superficially or as deeply as may suit the reader's interests.

In the acknowledgments section I religiously include all those in the private and governmental sector who contributed to the information, unless they prefer to remain anonymous. Any reader can understand the extent of the information network and can appreciate the likelihood that significant information was overlooked or not. If someone in government is doing a critical review of the study, the acknowledgments facilitate such review. The communications specialist or project coordinator must check with the subconsultants about people or groups contacted, no matter the number involved. Frequently, in the historical analysis alone, 10 individuals and five to 10 groups may have been contacted. In this section, the technical in-house staff who prepared special sections of the report should also be identified. It is helpful to those vetting the report.

The general page layout should follow a consistent style, indentation, titles and headings, subheadings, and margins. Tables, figures, maps, footnotes, and reference citations should also be consistent in terms of typing, lettering, capitalization, etc. There are numerous style manuals that can prove helpful in this regard, or the communications specialist can set up an in-house guide or program a word processor to facilitate text preparation, typing, and drafting.

As emphasized under the section on the communications specialist's role, the selection of language appropriate to the target audience is crucial. Since most consultants and clients have postsecondary degrees, the general level of sophistication can be much higher than the version for the general public and the press.

All unfamiliar technical terms should be footnoted or explained briefly in the text. Abbreviations or symbols are troublesome—e.g., DBH (tree diameter at breast height, $4\frac{1}{2}$ feet above the ground), CMP (corrugated metal pipe), or $N\text{-}NO_3$ (nitrate nitrogen)—to someone not closely associated with the particular discipline. Wherever a government body and professional association set biological standards that are of significance to human health or to plant and animal populations, footnoting comments on these standards helps readers evaluate the significance of the technical data. For example, 1 ml/L of $N-NO_3$ or more is considered significant in terms of adversely affecting water quality, or surface water temperatures exceeding 68°F or 21°C characteristically eliminate a trout fishery. Handled in this open and clear manner, technical terms do not confound the uninitiated.

Appendices help streamline a text, especially for repetitive tables or figures. Quantitative data on ecology of forest stands, water quality, and bird species observed at various times tend to clutter up the text, creating "visual friction" for a general reader. Such data can be reduced by using photography or by photocopy machine; thus the overall bulk of the report is reduced without sacrificing content.

Résumés of the key professionals are usually included in the appendix, especially if the report is to go to a hearing body or if it is circulated for comment to government agencies.

If a summary report is prepared as a separate freestanding document, it should contain key maps, figures, and tables identical to those in the unabridged report. These reports, prepared for general public distribution, may cause confusion and loss of credibility if the technical data are altered from the original or appear altered to downplay significant findings.

Graphics for Reports

Careful selection of appropriate line drawings, figures, and photographs tremendously improves the readability of the report, facilitates discussion of significant information, and improves retention of concepts. For reports of 15 to 25 pages, it is perhaps not as crucial, but over this length it is a definite advantage to include graphic materials. In drafting, size of lettering and any fine detail have to survive a required size or photo reduction. Maps are particularly troublesome, since contour lines, road names, numbers, etc. may be obliterated when reduced by more than one-half. If special photographic reduction is required, it is wiser to check with a commercial print shop before arriving at the shop with a series of maps requiring reduction and separate printing or blueprinting.

Fold-in maps, which are bound with the text, do not easily separate from the report. However, reduction to a maximum fold-in size may not be practical. Alternatively, maps can be folded and placed in a rear pocket or put into a separate folder. All separate maps should be numbered and have the name of the project and the consultant on them so that when they become separated from the report, they can be reunited easily. Using separate maps should be avoided. There is a special place in heaven reserved for graphics specialists who bind maps into reports, and there is a special place in hell for those who fold large maps and place them in the rear pocket.

The use of separate transparent maps to be overlaid to produce composite black-gray-white constraint and opportunity maps can be troublesome. "High," "medium," and "low" value areas must be colored or shaded so that when they are overlaid, numerous grays do not appear as one or more black sections. Tonal density must be experimented with and adjusted depending on the number of consecutive overlays to be used at one time. Usually such transparancies are placed in a rear pocket to be used independently or bound into the report and carefully registered so that the same geographic areas are in juxtaposition. The same scale, obviously, is mandatory.

If in the final report the primary consultant uses certain maps and illustrations taken from the environmental subconsultants' report, the layout, the lettering, and the reduction have to be coordinated with the prime consultant before drafting begins. Usually the primary consultant prefers to work from draft copies submitted by the subconsultant and do the final drafting, reduction, and printing in house. The primary consultant provides subcontractors with the proper base maps and mylars. It is not unusual for landscape architectural and engineering firms to take over the total graphics workload for the project, relieving the environmental firm of this chore. Always, a clean final copy, as opposed to submitting rough maps and figures to a prime consultant, facilitates retrieval or revision of the report. The maps produced by the prime consultant's staff should always be proofread by the subconsultant.

One other troublesome detail is the matter of map scale or a common base map for all the subconsultants to use. Often the numerous maps on a project (often at different scales) unfortunately come from different base maps. For this reason, they may register poorly for producing overlays. Street names, roads, rivers, villages, and city boundaries can differ between various maps, making it difficult to locate the same point or area. As well, metric maps and maps in imperial units have different contours. Agreeing early in the project what is the most appropriate base map avoids these difficulties. Using this same base map for as many maps as possible in the report is wise. Finally, place names used on the maps should agree with those used in the text, and vice versa; vernacular names used interchangeably with more formal and legal names confuse the reader unfamiliar with the area.

Graphics for Public Presentation

Graphics used for public presentation present a special problem, since a great deal of time and cost are required to prepare a public display. To prepare a 3 by 3 foot or 1 meter square display may take 3 to 5 person-days of time for middle-priced staff. In most cases, such public displays in color further complicate the task of layout and harmony.

The type of presentation determines the graphics approach. For example, if the display is for a drop-in center where people circulate around the exhibit, the work should be displayed on panels ranging in size from 3 by 3 feet (1 by 1 meter) to 4 by 6 feet (1.5 by 2 meters). Lettering of major features should be at least 1/2 inch or 1 cm in height; other map details can be fine, since the public examines the display at close range.

If any discussion of the display material is on a one-to-one basis, the consultant can point directly to areas of interest and key landscape features. If the presentation is to a seated audience of 20 to 200 people, the consultant can use either overhead transparencies, color slides, black-and-white slides, or panels having much larger lettering and simplified map symbols and features. The size of lettering is determined by the distance the audience is seated from the display and the projection distance. Commonly, when multiple presentations are made over many weeks in various settings, each setting has individual problems of darkness, lighting, and distance. Such variability requires advanced warning to the graphics coordinator.

Overhead transparancies, if they are to be used singly or in combination, can be handled in two ways. The simplest approach is to separate them and combine them according to a predetermined order. However, in a freewheeling public discussion, they can be mixed up, causing the audience to lose the thread of the discussion. With up to five overlays, avoiding this confusion is possible if the overlays are taped together at their edges forming a cross—four "arms" around a central key map overlay. Singly or in combination, the overlays can be flipped into place from left, right, top, or bottom, as the tape hinge holds them firmly in alignment. If the discussion requires redisplaying the overlays or using different combinations to illustrate different spatial interactions, this can easily be done.

Computer maps are difficult to use effectively in public meetings, because the varying intensity of features may require unfamiliar symbols to be used. The resulting "texture" can best be appreciated at a particular distance from the map, which may not be under the discussion leader's control. Orientation is also difficult unless roads and other key features are defined. Conceptually, the idea of machine-processing spatial features on a map seems to cause a psychological block for many people. Whatever the reasons, some types of computer maps seem to retard discussion and the free flow of ideas and antagonize or intimidate some people. In my judgment computer maps of resource features should only be attempted under carefully controlled conditions in open public hearings and be used sparingly before technical audiences. Their place belongs in the back room, where technical staff experienced in computer concepts and graphics are not inhibited from getting the message because of the medium.

Admittedly, if computer graphics are needed to synthesize mathematically resource features, computerizing is the only practical way to do this. However, one way out of this communication maze in public meetings is to convert such computer printouts (of an algorithm) into a conventional map form.

When presenting general ecological findings relating to a design concept in a public meeting, I plan for no more than a 20-minute presentation. Hence, the graphics to support such a presentation are best kept to a minimum. For example, in a presentation to a dozen decision makers regarding design of a regional park, I prepared a 2 by 3 foot or 0.6 by 0.9 meter set of transparent panels overlaying a base map. When the presentation began, I first discussed the general design concept of more intensive park activities to the north of a stream and passive activities south of this stream. The social advantages of this spatial arrangement were obvious, because the existing urban area abutted the park to the north. However, the environmental quality rationales were not easily grasped; this was the purpose in preparing the overlays.

The first overlay showed suitable (light) and unsuitable (dark) soils to support year-round active and passive recreation uses (a "high" and "low" or two-tier value system); the dark areas were organic soils. The second overlay showed forest areas (dark) poorly suited ecologically for heavy recreation use. The last overlay showed noise intrusion zones, suggested hiking trails, parking lots, and playing fields. Together these simple overlays conveyed the message that from a noise, soils, and vegetation point of view, intensive activities were reasonably compatible with existing natural resources north of the creek, while a selected range of passive activities would be compatible in the upland forest areas and in the organic soil pockets. This three-map sequence incorporated the ABC map system (see Chapter 3).

This simple graphic presentation, color slides, and figures supporting the rationale for a buried storm sewer versus a ditching of the creek resolved the spatial conflicts arising between engineering necessities and park aesthetics. This brief presentation was sufficient to open up 2 hours of spirited discussion of alternatives.

Report Production

Because of time, cost, and variation in the number of copies needed, obtaining multiple copies has to be considered for each project. In all cases, the client should specify in the contract the number of copies required, usually between 10 and 25. It is important to reach agreement on this issue before striking a project budget. As outlined below, if 35 rather than 10 copies are required, the additional cost of a long, well-illustrated report may run to many hundreds or even thousands of dollars. For 25 copies, photocopying is the simplest, most rapid, and least costly method.

Maps can be copied on a blueprint machine, trimmed, folded, and inserted in their proper place. If more than 25 copies are required, offset printing or an equivalent may be no more expensive and even be cheaper if additional runs are requested.

Costs of Report Preparation

Report preparation requires careful budgetary planning in addition to the timing requirements discussed earlier. It is surprisingly expensive to produce a tightly edited and well-illustrated report. Table 6-1 is an example of a budgetary breakdown for 25 copies of a 200-page report to illustrate this point.

Allowing approximately 10 to 15 percent of total budget is a reasonable working figure for editorial, graphics, typing, and copying services, assuming the communications specialist carries out the major functions of linguistic interpretation, proofreading, and supervision of graphic preparation. If the project coordinator and/or project manager does many of these "Joe" jobs, it is much more expensive, even though the cost is hidden on a different line of the budget.

If the client objects to 10 to 15 percent or so of the budget being allocated to the preparation of the final report, with a little imagination many of these costs can be dispersed throughout the budget. The editor becomes a "project assistant," for example, and graphic costs are allotted to each subconsultant. In any case, these costs are relatively inflexible no matter how they appear in the budget. If they are underbudgeted, either the quality of the project suffers, the gross profit is cut, or the environmental manager is working for less than usual daily charge-out rates.

Costs of Report Presentation to Hearings and Public Presentations

Generalization about costs for presentation to large audiences is quite difficult because of the variable nature of what may be required. For a slide presentation using graphics and maps prepared for the final report, costs are only a few dollars and are absorbed easily in the budget (Table 6-1). However, if panels and overhead transparencies are prepared for a drop-in center display or for presentation to an audience of 50 to 250 people, the salary costs to the project, as mentioned earlier in designing and preparing the graphics, can run in the range of $1,000 to $2,000.

In the widely used film produced by Ian McHarg in the 1970s, "Multiply and Subdue the Earth," some idea of the complexity of environmental graphic display can be grasped in the section of the film

Table 6-1. Costs of communications specialist's and graphic specialist's services for 25 copies of a 200-page report.

		Communications and graphics costs	Project cost
A.	Total fieldwork budget for subconsultants (soils, vegetation, water quality, etc.) including draft reports		$15,000
B.	Project supervision (salary)		5,000
C.	Travel, meals, miscellaneous (15% of A and B)		3,000
D.	Editorial salary (10 days @ $200/day)[a]	2,000	
E.	Graphics preparation: maps (5), illustrations (8), figures (5)	600	
F.	Map blueprint reproduction (5 maps × 25 copies–@ 75¢)	94	
G.	Typing ($1.00/page × 200 × 3 drafts)	600	
H.	Photocopying charges (25 copies × 200 pages @ 20¢/page)	100	
I.	Covers (color and illustrated) and backs	50	
J.	Spiral binding (25 copies @ $2.00)	50	
	Subtotal, communication and graphic specialists' services (D to J inclusive)	$ 3,494	
	Grand Total	$26,494	
	Percent communications specialist's and graphic services budget (D to J inclusive) to total budget	13.2%	

[a] This includes a markup of times 2 on salary, i.e. $100/day × 2.

where he presents his findings on colored panels, approximately 5 by 8 feet or 1.5 by 2.4 meters. Panels of that size, in color, require many weeks of drafting, not to mention color selection to achieve a harmonious effect for the overlays. They likely require much closer to 25 percent of the total budget than the 13 percent budget figure used in Table 6.1.

In all my work I have never been called on to produce such complex and costly graphics; a landscape architectural or architectural firm, if required to produce complex presentation material, would use draft material provided by the environmental firm. Graphics done by environmental firms characteristically are more straightforward black-and-white illustrations, figures, overlays, and maps. Occasionally I have glued computer run land use analyses together to make 3 by 4 foot or 9 by 1.2 meter map sheets, but these are not complex graphics in the sense of requiring artistic skill and large amounts of time (other than the computer time). If they were colored, however, some additional manpower costs would be generated.

This chapter has outlined the importance of considering the report phase of a project. It is important in preparing a budget, coordinating the work of subconsultants, and planning the time required for editing, graphics preparation, and printing. When many years have passed, the client and other professionals forget the cost and time problems. They look at the final report and judge the quality of your work: it had better "pass muster"!

7

Savings from and Costs of Environmental Management

Economic justification is one of the great difficulties I have found in marketing advice as environmental management consultants. Front-end costs (costs incurred at the beginning of a project prior to any profit generation) have to be carefully controlled, especially by corporations, and to a lesser extent by governments, because of interest charges on such borrowed monies. If rates are high or uncertain, interest charges accumulate for many years before profit offsets losses. Since development companies cannot avoid some planning, all engineering costs, landscape architectural fees, and environmental management fees may be viewed as "frills" by the comptroller, the board of directors, or the city planning director.

Generally speaking, many private and crown corporations and government agencies today hire environmental expertise, because "sticky" environmental questions are being asked by approving agencies or because public action groups are aware of the need to explore issues of environmental quality before approval. This often puts the environmental manager in the position of working only at a conceptual level or at a reconnaissance level of investigation to keep the costs down and the client out of hot water instead of providing a more adequate level of ecosystem analysis. It is for these reasons that it is important to examine various aspects of the economic rationale for environmental studies.

Usually a client calls environmental managers only *after* he has gotten into some difficulty; the same situation applies to other professions. A client calls a lawyer to bail him out only after he has signed a contract and then run into trouble; a patient calls a physician only after all the home remedies have failed rather than use a preventive approach by having a regular checkup; many a livestock producer has telephoned the veterinarian only when an animal is down in the stall; and many of us endure a dental visit only when teeth are painful. So too can the environmental cost issue be viewed from this same perspective: the one of preventing a problem from occurring rather than correcting an environmental quality problem after it occurs.

To justify ecological analysis principally on the grounds of preventing costly errors in design or construction, it would be desirable to have some proof. Proof is examined in the next section. Before discussing this evidence, first it is useful to understand the professional context of savings and losses.

One possible economic disadvantage to design and engineering professionals is the nature of the development process itself. If a planning, landscape architectural, or engineering firm undertakes a preliminary study (such as a highway feasibility study or a design for a park), there is a strong implicit desire on its part to see the work move into the next, more detailed stage. This desire for construction comes from four sources. The first is the orientation of these design-engineering professions: they only stay in business by building facilities. Next is a psychological identification with the project that develops; third is the potential financial advantage to the firm in the additional work generated at the detailed design stage, a percentage of the construction contract, which is between 2 and 7 percent. Last comes from a desire to please the client who is probably committed to some type of construction. If no construction, even minimal construction, is recommended by an environmental manager because of an environmental constraint, a no-build position, irrespective of its technical merits, is not likely to receive attention and support. In fact it may well be viewed as professionally inappropriate to many on the engineering-design team.

The environmental manager may often be the only member of a team seriously interested in a no-build option or in a lesser scale of construction. Costs "saved" in terms of unnecessary construction are often appreciated more by the client, if they reach the client's ears, and by the public than by these other professions. I emphasize here that I am not accusing design and engineering professionals of unethical conduct; but, rather what is, considering the psychological and operational circumstances, a not unusual human or professional response, though often a subtle one. Since the environmental manager receives no percentage of the construction fee, unless environmental inspection is done, this places the manager in a unique position, a position of which he or she must be aware. The environmental manager must present "no-build" or reduced construction options carefully and document this position to maintain credibility and acceptability.

In discussing costs and savings resulting from ecological studies, it is helpful to be aware of these various professional cross-currents and conceptual realities. In this way, the environmental manager is alert to possible financial advantages that can accrue from ecological analysis and recommendations, while being aware that differing professional points of view, on some occasions, may be influenced subtly (or not so subtly) by the degree of potential psychological power (ego) or economic gain if the project goes forward.

Costs and Savings — Conceptual Issues

Of the six modes (Figs. 1-1 and 3-4), the new facility development mode is discussed first, as it lends itself for analysis of costs and savings. Since the other four modes are more subjective, such as regional development or policy formulation, no attempt is made to examine their costs and benefits. Discussion of the facility or corporate operation concludes this section.

The facility development mode, from my experience in ecological studies over the past 14 years, gives considerable knowledge as to why savings in facility development costs, sometimes substantial savings, accrue when ecoplanning is done. Though there are many reasons for this, nine reasons either have recurred or are significant, even though they may occur infrequently.

1. Ecosystem analysis done by an interdisciplinary science-oriented team is a thorough physical and biological site analysis. On the other hand, the usual site analysis done by a single design or engineering professional covers highlights but cannot hope to determine all of the major physical and biological structural properties and ecological processes at work, past and present. Such analyses of dynamic systems take months for trained scientists to measure and comprehend. Since the quality of the design process depends on the quality of the site analysis, it is simply a matter of the probability of error.

A design- or engineering-trained professional usually misses some crucial interactions—not unexpected in view of the specialized training. These errors in misconception are usually corrected by additional site construction costs. Corrective measures simply compound the cost to the client and may further benefit the designer or engineer on a fixed percentage of the additional construction fee. In economic terms, the question is framed in the following way: At what level of detailed site analysis does an (expensive) interdisciplinary study break even to justify such a more complex site analysis?

Characteristically, it is the earth science analysis that uncovers the greatest potential savings. Geologists see landscape interactions "horizontally," while geotechnical engineers see the world "vertically" through boreholes. Together their composite vision is improved. To achieve a break-even cost-benefit ratio and to have the greatest probability that the geosciences will discover savings, this component in the ecosystem evaluation should never be omitted.

2. In an interdisciplinary approach, which allows the physical, biological, social, design, and engineering fields to interact more cohesively in ecosystem analysis, the consulting environmental scientist provides a balanced view whenever construction alternatives are reviewed. Such a balanced view may not reduce costs but may increase the intangible social and economic values of the project for the same cost.

3. A considerable amount of time intervenes between project initiation and construction. With inflation and interest rates fluctuating between 5 to 10 percent a year, $5 million invested in land may cost up to half a million in interest charges per year. Every unnecessary day's delay in approval for development because of public hearings or lengthy government reviews prior to approvals, on this investment scale, costs approximately $1,500 a day. A tract of 500 acres worth $10,000 an acre represents an investment in land of the magnitude discussed here. If ecosystem analysis is done as part of the project planning and site design process, it provides *without delay* specific and accurate technical answers to specific questions. Hence the number of hearings or circulation procedures can be reduced by about 5 or 6 days. The cost of environmental analysis is recovered by an equivalent reduction in interest costs.*

In Ontario, as elsewhere, hundreds of acres of urban development have been delayed month after month because city councils, rightly enough, are concerned about questions of social and environmental quality that the design and engineering team were not well prepared to answer. The interest charges, in the form of options or notes to these lands, are to the advantage of the bank, trust company, and insurance company holding the securities but not to the development company or the home buyer who wants a home at a reasonable price and an environment of some assured quality.

The high and variable rate of inflation also adds another dimension to the cost-time relationship. When construction begins on part of a subdivision where all servicing has to be installed first, and if delays subsequently occur because of environmental issues raised by the new residents or the governmental approving agencies, construction delays costs considerably more. The fixed capital costs of servicing generate no revenue until houses are sold. Because of inflation, an issue too complex for this discussion, delays are expensive if the developer is working within a fixed budget and cannot easily pass on his costs (e.g., he has long-term leases negotiated at a fixed price on a building to be constructed). Ecological

---

*More details of 1977 environmental analysis costs for different kinds of projects, and prepared at a variety of scales, can be found in Dorney (1977); doubling these figures gives approximate costs in 1984.

analysis is an "insurance" shielding the client from delays necessitated by last-minute ecosystem study requirements. An example of this occurred in the construction delays at the Buffalo campus of the State University of New York when a young lad discovered a nesting *Buteo* hawk in the way of the bulldozer. Turning off the bulldozers because of public furor resulted in some high-priced nestlings (Moore, personal communication).

4. When the quality of a development is strengthened by the design based on a systematic environmental analysis, this quality has a positive market dimension. A development company that has earned a quality image has less trouble selling homes and building lots. For example, lots abutting on linear open space generate additional front foot value; lots backing onto wooded ravines sell at premiums. Depending on land value and engineering costs, it may be more economical to keep a valley natural, or partially natural, then to heavily engineer a channel on the assumption that more developable land ("lot yield") provides for more profit. An example of cost-benefit alternative treatment of an urban stream is in Table 7-1. In this example no opportunity costs or discounting calculations are included, as the time intervals for the three alternatives would be similar.

5. Environmentally sensitive design for housing done by clustering development is cost-effective to the developer, builder, and home owner. The principal savings are in length of streets and length of sewer/water/ energy and communication connections. These servicing lengths may approach a 40 percent reduction over the conventional grid subdivisions of equal housing density. On-site storage of storm water is often another cost-effective procedure in subdivision design, especially where downstream flooding is exacerbated.

6. There is little recognition by designers that natural systems, if they are (or can be made) compatible with various human uses, are self-organizing or self-maintaining, requiring little or no input or labor and materials. Any natural spaces that can be conserved, renovated, or replanted return, over the long term, handsome dividends in reduced maintenance budgets. The percentage of such areas can be from 5 to 25 percent on institutional, urban, and industrial lands. If crops (tree farming, vegetable gardens, or farm crops) can be harvested on such lands, small revenues are generated.

One example of this reduced maintenance costs, other than an experimental one published by myself (Dorney 1983), occurred on a south-central Ontario university campus. A third of an acre of natural hard maple-beech forest was to be cleared so it could be decorated by a university sign and "appropriately" landscaped. Assuming annual grounds maintenance costs at about $1,500 per acre per year on the campus, this one-third acre would theoretically reduce the maintenance budget by $500 per year. Capitalizing at 10 percent a year would give this

Table 7-1. Cost-benefit analysis of alternative treatments for an urban stream (MacNaughton 1973).

Item	Heavily engineered alternative (60-ft right-of-way for 1,000 ft)	Moderately engineered alternative (103-ft right-of-way for 1,000 ft)	Minimum engineered alternative (290-ft right-of-way for 1,000 ft)
Channel treatment	Straightened and concrete lined banks	Grass side slope 3:1	Natural floodplain with elevated back lots
Engineering costs	$45,000[a]	$14,000[a]	None
Potential net profit of buildable land recovered by floodplain filling of land worth $13,800/acre	$39,000[a]	$26,000[a]	$39,000[a]
Potential net profit (loss) on lots abutting creek valley over standard lot value	None	$10,000[b]	$40,000[c]
Benefit	$39,000	$36,000	$40,000
Cost	45,000	14,000	39,000
	(0.87)	(2.57)	(1.02)
Replacement cost/year	$ 2,250[d]	$ 700	—
Landscape maintenance costs at $1,000/acre/year	—	$ 2,365	$ 4,362

[a] Cost estimates based on 1968 prices.
[b] Assuming an inflated value of $5 a foot frontage for 40 lots.
[c] Assuming an inflated value of $20 a foot frontage for 40 50-foot-wide lots abutting the natural stream floodplain (2,000 feet of salable frontage).
[d] Calculated as 5% depreciation per year or 20 years total.

natural stand a standing crop value of $5,000. Parenthetically, it might be noted that this stand also provides outdoor educational opportunities and is a visual and noise buffer of the women's dormitory. The stand was saved from the bulldozer for educational reasons alone, although economic arguments could also have been used.

Another spatial analysis on the same campus suggested that up to 20 acres could be reforested with quality hardwoods and white pine. Using the same acreage cost figures, this would reduce the grounds budget by about 10 percent and could provide at maturity, in 60 to 100 years, some commercial timber.

7. Environmental assessments affect the bidding price of contractors. In one example in Ontario, an environmental assessment report for a pipeline right-of-way was used as a guide in preparing contractor's cost estimates for soil stripping, bedrock blasting, stream crossings, clearing, and grubbing. While a well-prepared environmental report should sharpen the bidding by reducing environmental uncertainty, a bonus for the pipeline company, a poorly prepared environmental report can mislead the contractor (see p. 142 for a detailed description of this event).

8. The aspect of intrinsic or intangible values, such as that derived by having wildlife in an urban area of well-designed subdivisions, is not easy to measure. However, in one comparative study between two developments in Waterloo, Ontario, one was designed solely by an engineer (Lakeshore 1279) and one by an engineering and landscape architecture team (Beechwood 1292). The net profit advantage in the more sensitively designed Beechwood subdivision was over $5,000 per acre in 1972 dollars (Table 7-2). These figures reveal the magnitude of profit that can accrue from quality design. Had more environmental science analysis been done for the Beechwood development, these differentials could have been increased further. The design differences can be seen in Figs. 7-1 and 7-2. The Beechwood systems are left in a completely natural state with no fill into the valley; Lakeshore Village uses a collector road looping through the development. Its valley lands were ditched; the amount of open space is considerably less.

9. Finally, regulatory costs for facility development are more than a conceptual issue and bear discussion. These costs are either costs of approval or costs of operation. Few comparative figures on approval costs are available. However, one interesting case is discussed by the Canadian National Energy Board (1984) where a $439 million pipeline cost $4,200,000 in approval and construction supervision costs, or 9.6 percent. Of this figure, 5.5 percent was judged to be unavoidable, and 4.1 percent was avoidable or could have been reduced. However, the review process uncovered an $83 million "error" due to oversizing the pipe; this savings reduced the cost of the proposed project by 15.9 percent. Whether or not the regulations and review process saved money is not easily determined. As well, it cannot be argued that the oversizing would not have raised all the rates of all users, resulting in potential inequities to some users.

Looking at regulatory costs another way, the winners and losers and the variable time periods involved make any economic analysis difficult. Inevitably value judgments must be applied. For a recent discussion of this aspect of costing values, the Law Reform Commission of Canada paper by Schrecker (1984) is important.

Table 7-2. Comparison of net profit breakdown for two registered plans.

	Beechwood	Lakeshore
Park acreage	5.529	0.05
Total acreage	43.75	14.55
Number of lots	138	71
Servicing costs[a]	$ 402,480	$119,200
Public utilities costs (60%)	26,000	18,000
Unit levy costs ($300/lot)	41,400	21,300
Survey and legal costs ($150/lot)	20,700	10,650
Recreation center and park development costs ($500/lot)	69,000[a]	—
Land costs ($6,000/acre)[b]	258,000	87,300
Total development costs	817,580	256,450
Total gross profit[c]	1,515,200	406,250
Total net profit	697,620	149,800
Net profit/acre	15,945	10,295

[a] Servicing, public utilities, and lot levy costs are actual figures obtained from the City of Waterloo.
[b] Land costs area assumed at $6,000/acre.
[c] Gross profit figure is a result of interviews with respective developers.

Corporate Facility Operation Mode

In the facility operation mode, the critical issue is risk assessment. Manufacturing of hazardous chemicals, inadvertant spills, and waste disposal present a special problem of costs and risks for both the worker and the public and have both short-term and long-term implications. Generally, the post–World War II manufacturing and disposal methods are inadequate. Enormous costs occurred for cleaning up the Love Canal waste dump (Niagara Falls): hydrogeological studies, human resettlement, property acquisition, and residents' compensation. This situation is an early example of what lies in the decade ahead. Because of Love Canal, Occidental Petroleum Corporation alone paid $5 million to $6 million in claims from "self-insurance," with the remainder in claims (unspecified) paid by insurance companies. A more recent example, the explosion and death resulting from the release of cyanide gas in Bhopal, India (December 1984), will cost Union Carbide not only large court payment and high overhead to undertake risk assessment of all its

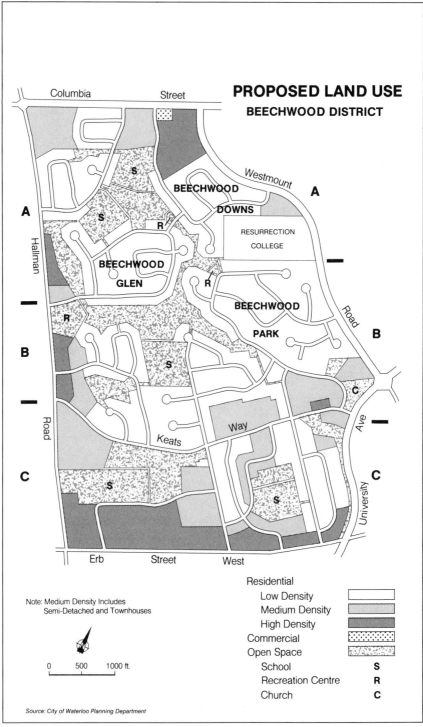

Figure 7-1. Proposed land use Beechwood District plan, from MacNaughton (1973).

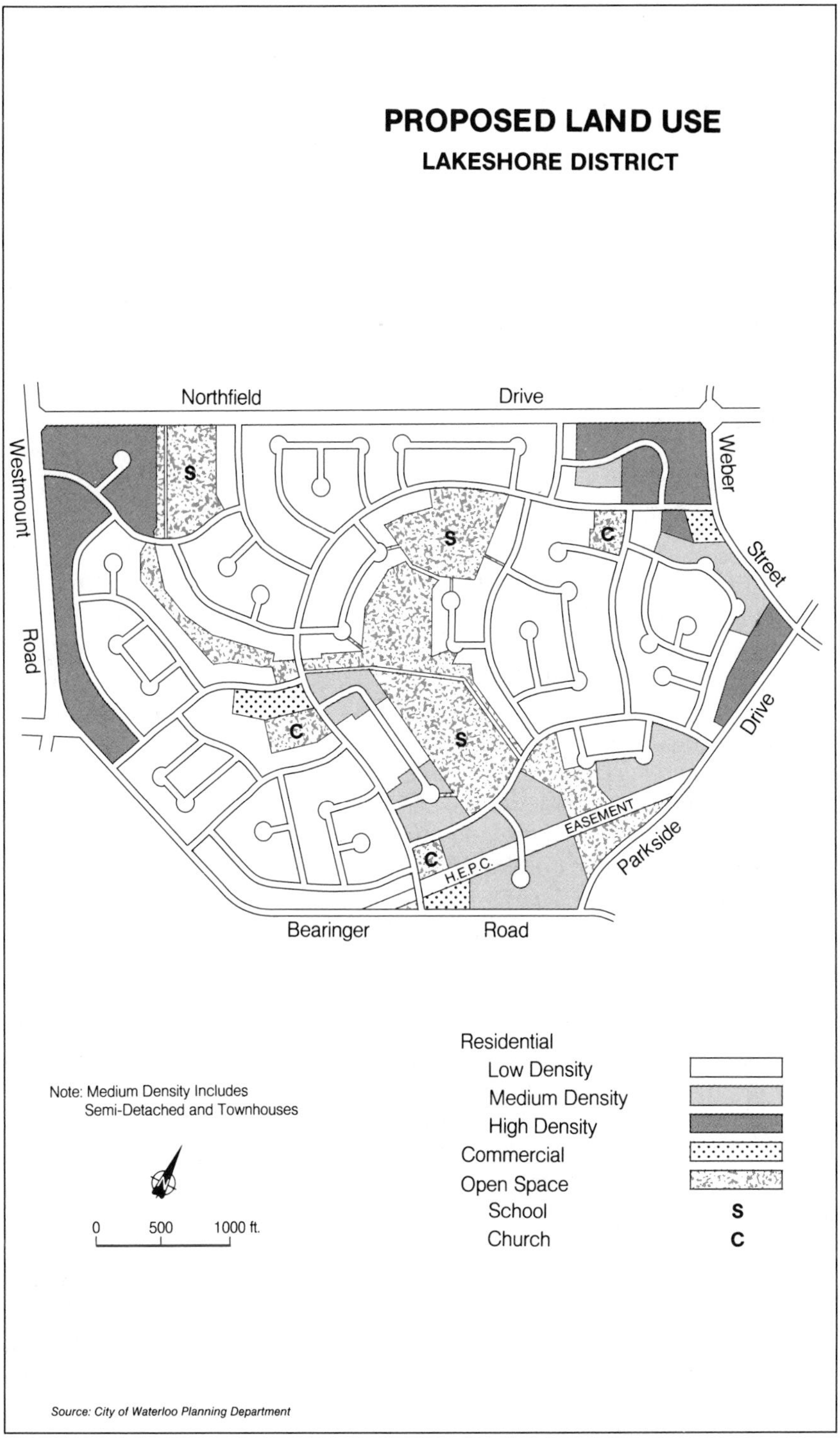

PROPOSED LAND USE

LAKESHORE DISTRICT

Northfield Drive

Westmount Road

Weber Street

Parkside Drive

S

S

C

C

S

C

H.E.P.C.

EASEMENT

Bearinger Road

Note: Medium Density Includes
Semi-Detached and Townhouses

0   500   1000 ft.

Residential
  Low Density
  Medium Density
  High Density
Commercial
Open Space
  School   **S**
  Church   **C**

Source: City of Waterloo Planning Department

Figure 7-2. Proposed land use Lakeshore District plan, from MacNaughton (1973).

facilities, but possibly large compensation payments to the 2,500 victims who died and the 60,000 hospitalized in India.

The benefits of reexamining the risks relating to the facility operation mode result in safer working conditions: fewer externalities or spills affecting areas outside the plants and reduced risk to nearby residents. The application of consistent risk assessment procedures, which are acceptable to regulatory bodies and which result in reasonable insurance rates, still lie ahead. If public perception of risk increases, decreased property values or forced relocation of facilities away from settled areas may occur. Should a panic or near panic situation ensue, resulting in stiff regulations, these and insurance rates may force plants to close, relocate, or modernize. The costs of products would rise dramatically, crop yields would decrease, and health costs would increase. Taken together these adjustments would ultimately be borne by all consumers who have a lower standard of living but live at a reduced level of risk.

Looking at environmental risk management in connection with the impact of toxic chemicals on public health, the field of ecotoxicology is beginning to reach the stage where large expenditures of public and private monies are occurring. Not only is the science of ecotoxicology in its infancy, but the potential for panic-induced regulatory changes is real. The costs and benefits, seen as a rational set of arguments on relative degrees of risk, could be overwhelmed by events, as they have unfolded in Love Canal and Bhopal.

Looking at the legal process associated with risk, the courts eventually must establish whether cause and effect are technically proved and what economic penalties or criminal charges are reasonable. As more risk-associated litigation is heard, the economics of environmental quality emerge in direct and indirect ways. In this context, again the recent Law Reform Commission study (Schrecker 1984) is worth reading; it presents a fascinating and thorough examination of the environmental hazard-economics-legal paradigm. Any evolving body of knowledge or procedures for establishing cause-and-effect evidence must be able to withstand the rigors of a legal test.

Examples of Potential Savings Derived from Specific Ecoplanning Studies

It would be naive to suggest that ecosystem analysis, when done for the purposes of improving the fit between nature and a desired land use, guarantees in every case a cost savings to the government agency or private company undertaking the work. The previous discussion suggested some of the broad conceptual issues. It is not uncommon to identify substantial probable or potential cost savings in development projects varying from a few thousand dollars to many millions of dollars.

The term "potential savings" needs qualification, because money not spent is not necessarily money saved. For example, if it is not spent on one aspect of a project, it could be spent in another. By "savings" in this context, I mean simply money that was not spent in an attempt to achieve an unrealistic or unattainable objective because of new insights reached through environmental studies. The following three case studies, derived from projects in which I have been involved, range in size from a four-county area of 2,400 square miles or 6,500 square kilometers to a campus of 1,000 acres or 400 hectares. They illustrate two of the environmental management modes (Fig. 1-1): facility development mode (dam development and a new university campus) and the urban and regional development mode (new town planning). A range of the economic costs and benefits are also presented.

New-Facility Development Mode

Dam Development

In the Waterloo regional municipality and parts of adjacent Wellington County (Ontario), a land use study analyzed numerous resource management and ecological issues, and a range of policy recommendations for land use were formulated (Dorney and George 1970). However, a major agricultural, recreation, and water resource conflict arose over the proposal to dam the Grand River at West Montrose (about 15 miles or 24 kilometers north of Kitchener-Waterloo; see Fig. 7-3). This dam was somewhat of a borderline case—marginally justifiable for flood control, low flow augmentation in summer, and recreational enhancement. The case for the dam was based on two cost-benefit analyses. However, at that time no environmental impact analysis was made, nor were there alternative development proposals made, such as varying dam heights and operational characteristics. The obvious losses in agricultural production (high capability lands), disruption to the Mennonite agricultural community, possible flooding and damage to the geologically unique Elora Gorge, and impact of increased recreation use on the fragile white cedar woodlands in an existing park (shallow soils over limestone bedrock) were not identified.

These issues of environmental quality of dam construction were brought to the fore when the ecological study (Dorney and George 1970) was presented to the then County Council. Its presentation resulted in the council's forming a Citizens' Environmental Advisory Committee. The committee, after reviewing the two conflicting cost-benefit analyses and the lack of environmental analysis, recommended unanimously in 1970 to the Ontario Treasury Board that a more thorough interdisciplinary appraisal be done and include social, biophysical, and hydrological

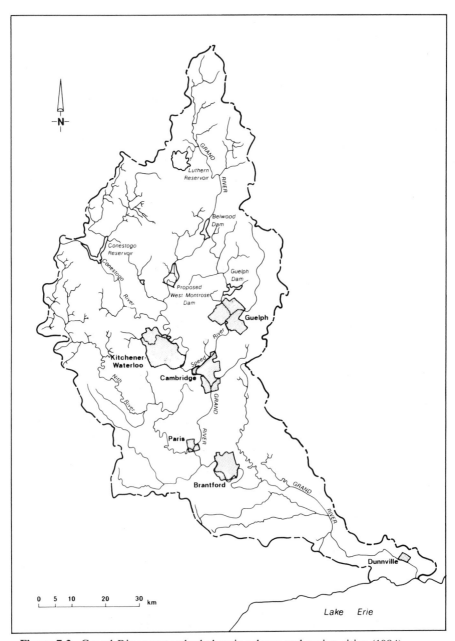

Figure 7-3. Grand River watershed showing dams and major cities (1984).

parameters. When the Treasury Board study was undertaken (Ontario Treasury Board 1971), it was apparent that the dam would not perform hydrologically as it was expected to do.

The issue was, in part, one of scale. The Environmental Advisory Committee and the proponent were conceptualizing the problem as a dam with some surrounding farmland, a problem at the ecosection scale (Fig. 3-1). The Treasury Board study team, which took a systems design analysis approach, viewed the problem at two scales—the ecodistrict and the ecosection. When the more generalized concerns at an ecodistrict scale were studied, it was apparent that the existing configuration of dams and tributaries was a concern (Fig. 7-3). Two major tributaries, the Nith River and the Speed River, having at that time no flood control dams, are upstream of settlements having flood problems (Cambridge, Paris, and Brantford). Since the proposed dam at West Montrose would have been the third dam on the same Grand River section, protection from a local convectional summer storm would be better achieved by first securing flood storage on the two tributaries. This was done in the mid 1970's by building a dam first on the Speed River and delaying the West Montrose dam.

It would be presumptuous to attribute the Ontario Treasury Board study, which may have cost on the order of $85,000, solely to pressure derived from the Environmental Advisory Committee, since the committee was just one of many groups and individuals who believed that a more thorough water management cost-benefit study was needed. Nevertheless, the fact that the committee quickly agreed and responded is a reflection of the credibility and value of the biophysical mapping previously done (Dorney and George 1970). Since the costs of this dam would have been in the tens of millions of dollars, it would have been an unjustified expenditure.

In the late 1970s a second comprehensive study on this same watershed was done. This more recent study contains considerable cost-benefit analyses and considers alternatives, including provision of improved drinking water supplies (Grand River Implementation Committee 1982). Subsequently, significant gravel and stone deposits were identified in the West Montrose reservoir area (Ministry of Natural Resources 1981). Extraction of these deposits alone, prior to any dam construction, would result in a deeper reservoir and in considerable revenues to offset construction costs. In addition, the now regional government promoted household water-saving devices, a low-cost solution in contrast to providing more potable water through dam construction and/or a Lake Erie water supply pipeline.

As this example points out, it is a sad commentary, and a not unusual situation, that multi-million-dollar construction projects often seem to receive a token evaluation. If these environmental management studies and interventions cannot always play David alone, at least they can provide some of the ammunition for a David to use. This particular dam project was only months or weeks from final approval before the 1970 Treasury Board review was undertaken.

As other major facility development efforts emerge in the future, the availability of resource mapping, the presence of environmental advisory committees, and the requirements for environmental assessment will challenge foolishly conceived projects, either altering their design or preventing their implementation. The dollar value of this response is difficult to measure, but as this example demonstrates, it can be substantial.

A New University Campus—Utica, New York

Cost savings in this 1,000-acre or 400-hectare project site, which might be attributed at least in some measure to the biophysical analysis, fall into four categories (for additional details see Dorney 1973):

1. Additional costs to construct buildings, roads, and parking lots in a compact clay till and unstable shale bedrock area rather than concentrated on a sandy beachline.
2. Additional costs regarding the replacement of natural sensitive forest stands indirectly impacted by construction.
3. Reduced maintenance costs for lands held in reserve for future construction.
4. Hazards to buildings of naturally occurring gas.

These savings can be viewed from the perspective of the $28,000 consulting fee for the environmental study. The major design issue was either building the main campus complex on the upper Mohawk Valley (till and shale) slopes providing a scenic panorama as opposed to utilizing a lower sandy river terrace which was more stable geologically. This dilemma was resolved partially on the basis of engineering suitability: the upper slopes are a compact till subject to drainage problems; the middle slopes are underlaid by shale, which presented difficulties for foundation support; and the lower slopes with a poor view are covered by a well-drained sand-and-gravel river terrace formed at the time when the Mohawk River carried much of the glacial melt waters from the Great Lakes to the Hudson River. This terrace was an easy zone for construction. The exact financial costs between these three sites were never computed, but some concept of the magnitude of the differences is apparent: the compact tills require "ripping" by heavy machinery for excavation; the shales, subject to rapid breakdown in less than 24 hours when exposed to the atmosphere, require special "rapid" construction techniques; but the sand-and-gravel terrace is easily worked at any season by machinery. A nearby hospital that had been built on the shale had considerable differential foundation settlement and building damage, apparently resulting from improper safeguards during construction.

New York State road construction standards require for these tills a base of 3 to 4 feet of sand and gravel over them to provide drainage and prevent spring breakup of road surfaces. The cost of this base material is considerable when the linear feet of road surface in a large university campus and the area dedicated to parking lots are considered. Again it is hard to put a precise dollar value on these construction penalties, but assuming a 3-foot or 1-meter fill requirement, a road network of three-quarters of a mile or 1.2 kilometers of two lanes, 30 feet or 9 meters wide, and 30 acres or 12 hectares of parking area, the total cubic yards of sand/gravel required would have been about 160,000, or 122,330 cubic meters. The cost of the amount of sand and gravel alone exceeds the entire cost of the environmental study by 17 times.

Regarding the second issue, there was a severe potential penalty because a mature hemlock stand near the building complex would have been killed by the design which proposed a complex of buildings to be "wrapped around" the wooded area preserved ostensibly for amenity purposes. These penalties would have been of two types: the landscaping costs for removal of dead trees and replanting, and soil erosion and bank stability problems resulting from the loss of the forest canopy. Since about 20 acres or 8 hectares would have been affected, removal and replanting with sapling-size trees (2 to 3 inches or 5 to 7.5 cm in diameter at a stocking density of 50 trees per acre at approximately $50 per tree) would come to around $50,000. This assumes salvage value of the mature hemlock would equal its cutting costs. Again, this hypothetical amount exceeds the value of the environmental study by about two times.

Soil erosion control costs are hard to estimate. Because of the gradient and runoff generated by the building, road, and parking lot surfaces, concrete or corrugated pipe drains would be required, possibly in connection with energy-dissipating weirs, to control bank erosion and protect bank stability. The gentle slopes and inherent permeability of the sand-and-gravel terrace require less drainage control. It would be quite arbitrary to put comparative cost estimates on such drainage structures, but similar engineering work on the Toronto Zoo site has cost tens of thousands of dollars, possibly approaching or, again, exceeding in cost penalties the entire cost of an environmental study.

The third issue, that of reduced landscape maintenance costs, emerged from consideration of preserving bird habitat for upland plovers and shrikes, allowing resident farmers to maintain viable-size operating farm units and maintaining open vistas as amenity resources. Since about 400 acres or 162 hectares was recommended by the environmental team to be retained in agricultural production (hay land, rough pasture), the cost-benefit assumptions depend on the degree of artificial landscaping that could have been applied to this large acreage and on the rental value of it to farmers. Assuming required mowing twice per year on this 400 acres, it could cost the university approximately $20,000 per year. Rental value of the area for agricultural purposes, on the other hand, might be equal to or as low as one-fourth this amount, depending on the local demand for hay and pastureland. In addition, such a rental arrangement would maintain a viable farm unit adjacent to the campus, a benefit to the farmer and the tax base of the township. Since these are annual costs, the long-term tangible and intangible value in retaining existing agricultural uses of the land demonstrate the strong long-term and short-term economic values in such land use recommendations made by the environmental team.

The fourth issue, the natural gas hazard, was most interesting. The historian working on the environmental team in his interviews with former farmers identified this construction hazard. Much to the surprise of all concerned, the site had been prospected for natural gas, and leakage into local well water was a common occurrence.

Overall, then, on this project there were a number of issues raised by the environmental study that had strong economic advantages when viewed from the perspective gained by a thorough ecosystem analysis. It is my belief that a factor of five to 10 times the ecological project cost could be attributed conservatively as savings that were the direct result of ecosystem analysis incorporated into the conceptual design. This is not an unreasonable way to view the costs and benefits, since an earlier, different plan had been prepared. This plan was scrapped because of the environmental analysis. As it would require the services of a professional estimator to compare the two plans, the figures presented here provide a rough approximation of the cost differentials.

Scientists usually do not appreciate the magnitude of the cost involved in construction penalties arising from poor knowledge of the geological and engineering properties of soils. Certainly the drama comes to a climax when structures fail and human lives are lost. Yet the benefits at the conceptual design stage resulting from a more thorough understanding of the biophysical environment appear to be substantial, and "preventive" penalties are avoided. This example affords a crude economic measure of the information gap between reality obtained by a thorough interdisciplinary, systems-oriented analysis and the reality as seen by designers and engineers using a reductionist approach. My experience demonstrates that considerable benefits are derived by sequencing the studies (biophysical first, then design) and merging the two perspectives.

New Town Planning in Ontario: Erin Mills and Townsend

Of the three new town projects in which I have been involved, two have evolved far enough into construction to give us some indication of possible construction savings resulting from biophysical analysis done in the planning phase. The first project, Erin Mills in Mississauga, Ontario, demonstrates the discovery of a geological constraint, which is reasonably clear-cut in terms of potential construction penalties. It relates to the appropriate land use designation in a complex valley system of about 1,000 acres or 405 hectares in size.

Originally, the new town planning team thought that detailed soils and surficial geology maps would not be useful, since other soil engineering reconnaissance had been done from aerial photos (4 inches to mile or 10 cm to 0.4 km) and the existing soils map (Fig. 5-1). Nonetheless, using research monies, we subcontracted a field investigation of soils and geomorphology at a scale of 1 inch to 400 feet or 2.5 cm to 122 m on an experimental basis (Fig. 5-2). The identification, mapping, and rating of some stratigraphic zones as having a high slippage potential caused not only surprise but some consternation among the planning team, since considerable housing, road, and service line construction had been planned for this entire area. Reasonably enough, a third engineering soils group was called in to do additional fine-scale bore-hole work; their analysis, more detailed than the previous engineering survey, verified the potential slippage difficulties.

The developer's purchase of adjacent tablelands allowed the planning team to redesign the whole block. The slippage zones were designed as open space. When the area is completely urbanized, an extremely attractive park area for the new town will be in place. The park, covering about 30 acres or 12 hectares, contains a magnificent stand of white pine and hemlock upwards of 80 or 100 years in age, associated with an attractive ravine system. The upland adjacent to the parkland contains high-density housing areas and thus achieves a dwelling unit density similar to what had been originally planned.

It is not easy to put a dollar figure on the value of identifying these potential slippage zones. However, the value of housing on, say, 100 acres that was to have been built on these zones would have represented an investment of the order of perhaps $12 million to $20 million. Using this figure as a simple and admittedly crude yardstick of either potential property losses or increased engineering costs to control slippage that might have been discovered at the detailed design stage gives an order of magnitude for the "savings" achieved from the detailed low-cost soil survey part of the ecosystem or biophysical analysis.

The aspect of caveat emptor now enters into the discussion. Such slope instabilities would not be considered legally as an "act of God," since a competent geologist or soil scientist would have, and in fact in this case study did, mapped them as hazardous areas. The legal liability to a planning team choosing to ignore such expert advice is a reasonably straightforward error of commission. Soils and geomorphology maps, subpoenaed by the courts if necessary, provide strong evidence for a plaintiff trying to collect damages for loss of property and human life. Had the mapping not been done for this new town, a professional error of omission could be claimed. It is noteworthy in this case study that similar slippage zones are a problem to housing along the east banks of another close-by river. These difficulties are well known locally and are often reported in the press.

The other aspect to this issue of slippage is that given sufficient funding, almost any limitation can be overcome (including in situ freezing of unstable soils during construction), if the client is willing to bear the cost. Either way, there are economic realities, construction constraints, and legal realities to this kind of an issue that can either penalize a development company in the courts or tarnish its public image. When a thorough physical and ecosystem analysis is done on a site, these realities can be brought into focus and then resolved by the company representatives, their lawyers, and the planning team.

The issue for the new town of Townsend was the identification, during the biophysical predevelopment survey, of an economic gypsum deposit buried in the middle of the new town site (Dorney 1977). By locating the town center to the south, this area is in a permanent agricultural buffer zone, enabling shaft and tunnel mining to be done at any time. The cost of geological studies to identify this valuable deposit worth hundreds of thousands of dollars was $2,500 (1975 Canadian dollars).

Examples of Costs of and Savings from Environmental Protection Activities

In theory, cost-benefit breakdown for environmental protection could be developed for each of the six modes shown in the right-hand section of Fig. 1-1. However, as the figure suggests, there are four principal areas of concern—the urban and regional operation mode, the facility development mode, the facility or corporate operation mode, and the facility decommissioning mode. These four modes will be discussed in turn. For a description of what is meant by "environmental inspection," which is part of environmental protection, the reader is directed to a report by Mutrie and Dorney (1981).

Cost and Benefits for the Urban and Regional Operation Mode

The budgeting for monitoring and for abatement and enforcement of environmental regulations is scrutinized by each jurisdiction, based on its perception of risk. Risks take the form of concerns related to public health from pollutants in food, water, air, and exposure in the workplace. The health penalties to individuals include direct injury, lost time at work, disability, early pensions, and a shortened life span. The worldwide public concern for toxic waste management demonstrates the critical and continued need for abatement and enforcement, monitoring, and research.

Property values are, of course, directly affected by pollution. Corrosion occurs, acid rain affects water quality and cottage owners' property values, and toxic dump sites are discovered near homes. There is no doubt of the continued importance of environmental protection in industrial countries. The United States Government, for example, spend $4.5 billion on pollution control and abatement in 1976 but reduced this to $2.2 billion in 1983. Whether this amount is too large or too small partially depends on political philosophy; it does demonstrate a significant scale of commitment. Costs borne by local initiatives of state, county, and city governments and the private sector give some sense of the importance of this field to environmental management. In Western Europe, expenditures for toxic waste management were $1.5 billion in 1984.

All expenditures are scrutinized by technical staff and economists and then approved by politicians who make hard tradeoffs. Hence, one would assume that considerable benefits must be apparent, although the precise relationship between expenditure and benefit is nonlinear. The large lawsuits regarding chemical spills and toxic waste dumps are one reason this mode is an expensive one for the government and the private sector—witness the Occidental Petroleum Corporation settlement of the Love Canal litigation.

A good deal of the research in this mode is on toxic chemicals and abatement technology. These are reductionist approaches. However, some of the monitoring can be systems based, such as sampling tissues of upper food chain fish and birds, as indicators of bioconcentration of pesticides or heavy metals. Acid rain, clearly a regional-scale, operational-type problem in North America and Europe, requires not only meteorological and emission model-building skills but considerable economic knowledge before abatement is agreed upon. The costs in property damage (buildings, autos) and in deterioration of fisheries and recreational sites, timber damages, poor human health, and possible alternative energy resources play a role in the decisions that are implemented. Also at this regional scale, public information and political negotiation and pressure are critical ingredients.

Costs and Benefits for New-Facility Development Mode

The costs of environmental supervision of the construction stage depend on the individual project, thus making generalizations difficult. To provide some feeling for the costs, a 25-mile or 40-kilometer pipeline construction project costing $25 million might have one full-time environmental inspector on the job for 5 months at $4,000 per month in labor and expenses. These direct costs can be viewed in the context of perceived or real benefits from having a full-time environmental inspector on the job. The benefits are in avoiding shutdowns by government inspectors for violating environmental pollution laws or environmental resource protection laws (such as unauthorized withdrawing of water from streams). In addition, documenting construction activities in environmentally sensitive areas avoids or controls future claims for damages by adjacent property owners.

The rapid growth of environmental inspection in the past 10 years (Mutrie and Dorney 1981) suggests that construction companies see merit in the concept, as it is perceived to be a general benefit—in spite of adding some cost to the project, estimated to be 0.1 percent of total budget. Since shutdown of construction is expensive, the environmental inspector has to evaluate the costs of replacing or renovating a resource feature opposed to the costs of delayed construction. These tradeoffs require sensitivity in implementation. In the final section, some specific pipeline cost data illustrate this point of protection versus renovation.

In another study, Nelson et al. (1981) demonstrate that pollution abatement costs for three large industrial developments (a refinery, steel mill, and electric generating plant) were 11.7 percent of total construction costs of $2.1 billion along with 6 percent in additional costs borne by the regulatory government agencies. Although these large expenditures met all environmental regulation requirements, the authors point out the difficulty in evaluating whether any real benefits are in line with these high costs and whether measuring the intangible social costs and benefits, such as employment generation, taxes generated to the local area, and the pros and cons of any lifestyle changes, are possible.

Costs and Benefits for Facility Operation Mode

Many corporations, especially in the energy, transportation, petrochemical, and chemical sector, have environmental quality departments or sections. Their function is to monitor emissions, prevent spills, supervise cleanup of accidental spills, and organize worker education programs. The key personnel are usually chemical or civil engineers familiar with operation of the facility or facilities. Rarely are biologists, geographers, or social scientists part of the staff.

Although no cost figures on the activities of these groups are available, to my knowledge, the few people in these sections suggest costs are low. The benefits perceived by top management are reducing government prosecution or shutdowns when violations of pollution orders occur: for example, for a railway company when a train derailment causes a toxic spill.

In the case of accidents, fire and police personnel may not be aware of legal notification requirements. Many environmental departments of government require notification of spills so they can counter any impacts on drinking water supplies, water supplies for livestock, etc. Should the agency or company causing the spill not notify the government, fines can be levied. Having environmental management groups in house assists in promptly notifying the company of the regulatory restrictions to minimize fines and/or legal claims for damages.

Costs and Benefits for Old-Facility Decommissioning Mode

With the decommissioning of the Three Mile Island reactor and the Chernobyl disaster, certainly a visible example of a major facility breakdown requiring decommissioning, everyone is learning how to undertake this operation. The lessons are costly. With unproven or new methods requiring some experimentation, such projects are expensive, but the benefits accruing are generally to the public at large, not to the corporation or government. If costs are borne from general corporate revenues, this may place the utility in danger of bankruptcy.

Other facilities requiring decommissioning, such as old factories, do not have an environmental management component. However, old industries using radium and heavy metals are hazardous to demolish. Tanneries, for example, used heavy metals in processing leather, and many watch factories used radium in dials. Rubble from these sites, dust raised during demolition, and excavation of soil adjacent to the building can present health hazards to local residents and workers.

Summary of Costs and Benefits of Ecosystem Analysis

As the examples used above indicate, based on the spectrum of consulting practice and academic studies, there are many conceptual issues of costs and benefits. In many situations, ecosystem analysis identifies particular courses of action that may be less expensive to follow or may generate higher revenues than alternatives. Often, these courses of action may be overlooked and discarded out of hand, demonstrating the utility of using a systematic multidisciplinary approach, combining both a reductionist and a systems approach.

However, there is another possible interpretation some might draw from these examples—that is, with a little more sensitivity to the environmental issues, a process called "scoping" or problem definition, a specialist in geology, history, forestry, or hydrology may be all that is actually needed to reduce the extent of the analysis and thereby the planning costs. Scoping may enhance savings, but from another perspective, what problems can arise when the analysis is simplified at the outset? This aspect then clearly deserves some elaboration.

The difficulty in scoping is that it requires in a single person or a few people with a wide range of experience in a wide range of fields to cover a wide range of potential issues. Such key thinkers in environmental matters, called by some systems ecologists (see Odum 1983), are not numerous. Coming from fields such as architecture, engineering, or economics, these thinkers have spent years working in a reductionist mode; a systems mode of understanding an ecosystem is not easily grafted onto a reductionist one. Rather, it is the constant interaction with systems professionals in engineering, natural sciences, and social sciences that brings new insights to the senior professional taking on organizer and synthesizer roles. Once an individual's source of systems knowledge specific to certain kinds of problems becomes dated, knowledge and credibility suffer. Put another way, it is often the free-flowing discussion from an interdisciplinary group that causes deep insights suddenly to emerge. This human intellectual synergy needs the stimulating climate brought about by a group process. The technical principles outlined in Chapter 2 (or something similar) can form the basis for scoping.

My experience demonstrates that too narrow a problem definition at the outset of a project results in enormous cost overruns later. This is demonstrated in a report by Loucks et al. (1982) where excess costs of up to nearly $1 million were incurred in megaprojects by initially inappropriate environmental management procedures. In another perspective, the relationship between increments of environmental management costs and increments of total project costs is nonlinear. Determining the cutoff points is never easy.

Notwithstanding the value I perceive in having a reasonably thorough ecosystem analysis done by as broad an interdisciplinary group as possible, there are instances where single environmental disciplines contribute to a better economic solution than one made by a planner, architect, landscape architect, or engineer acting alone. I would rank these discipline contributions in descending economic terms:

1. geology (including hydrogeology and soil science)
2. hydrology
3. forestry
4. limnology and surface water quality
5. historical ecology
6. climatology
7. aesthetics

Wildlife population analysis has not, in my experience, ever generated economic advantages to a development proposal. This does not mean, however, that it should be omitted in an ecosystem analysis, even though wildlife resources defy easy conversion in economic-design terms. This is particularly true of rare or endangered species whose market value is low unless tourism can be demonstrated—for example, people traveling to Scanlon's Lagoon in Mexico sight to the gray whales.

Epidemiological analysis, theoretically, could make important economic savings where disease transmission through poor landscape design encourages the flow of diseases from natural nidi or source areas to humans, thereby entailing more costly control techniques. Such situations are more likely to be found in tropical areas, but they may also occur in more temperate regions. For example, plague, rabies, encephalitis viruses, and rickettsias are common in Europe and North America. Encouraging certain habitat types where alternate sylvatic hosts can reach high populations implicitly carries certain hazards and cost penalties for control.

Environmental mediation (neutral third-party mediator), negotiated development, or informal conferences at the prehearing stage are other mechanisms to reduce litigation costs (Horte 1983; Nelson 1982). Mediation should be encouraged, if both parties have room for flexibility and the mandate to negotiate. If such a mandate is uncertain, or if one party is perceived as having all the power or strength, serious negotiation is not possible.

Compensation for risk is another device being tried where communities may be coaxed into accepting money to reduce impacts of development rather than go through costly hearings having an uncertain outcome (Pushchak and Burton 1983). In Ontario, a process similar to this was done near the Darlington Nuclear Generating Station; in this situation the local community received monies to mitigate probable social impacts.

Taken as a whole, an ecosystem analysis comprising ABC components (Fig. 1-3; Dorney 1973) can result in demonstrable cost savings on many projects. Ecosystem analyses uncover problems and risks that would have been or were overlooked in the conventional single-discipline or reductionist analysis. Although any calculation of benefit-to-cost ratios from such interdisciplinary ecosystem studies must be used with great care in discussion with clients, based on my experience, the ratio can range from zero to as high as 20:1. This 20:1 ratio should not be oversold or used carelessly, as it will not always be apparent. I believe the following conclusions are reasonable:

1. The environmental sciences singly and collectively through ecosystem analysis can often justify their inclusion in addressing urban or regional problems on tangible economic grounds and on the grounds of improving design solutions which in turn can lead to other intangible short-term or long-term social and economic benefits.
2. Professional designers and engineers may not be aware of the depth and sophistication available to them, at reasonable cost, in the various

natural science and social science fields; consequently, overall professional performance can be improved when expert advice from these fields is brought into the planning arena at one point or the other.

3. Government regulations regarding environmental assessment and design may *decrease* the overall development costs of projects and provide an insurance against major miscalculations and blunders because of conceptual blinders of many professionals.

4. Executives may not be able, in spite of difficulties in the precision of an economic study, to claim that costs for environmental protection (planning and design) exceed benefits (Thompson 1980). The reverse has been demonstrated in some of the examples cited above.[*]

5. Environmental managers no longer have to justify their involvement in an urban and regional development project or in a facility development project as interlopers; they can advance some positive economic arguments for their involvement. This in no way undercuts the more conventional ethical arguments of how humans and nature should interact; rather it should reinforce them.

6. Executives claiming that environmental studies are producing uncontrolled costs and slowing down economic development certainly have a point (Loucks et al. 1982), but the counterargument of savings generated by utilizing environmental management more effectively can be made. In this context, it is not what you do; it is the skills brought to bear in a somewhat unfamiliar format and the expediency achieved that make the effort successful or not.

Costs of Ecoplanning Analysis as a Proportion of Total Planning Costs

In budgeting for ecological studies, the costs of each of the case studies described are small. Because of their quasi-experimental nature, some of the studies discussed above were paid under government grants that usually do not allow any or allow only a partial professional salary to be charged. Nonetheless, it is easy to estimate, from the amount of professional time actually spent, what would be an adequate budget.

Government fiscal administrators and potential clients in engineering or design-oriented firms normally have to strike a total budget proportioned between their firm's staff requirements and that of their subconsultants in ecology, economics, sociology, etc. For this reason, it might be helpful to present my experience in the hope that realism can be brought to bear on

---

*Similarly in the Nanticoke industrial development study, Nelson et al. (1981) were able to identify environmental protection costs as 11.7 percent of total project costs. They believed the benefits were essentially in the "eye of the beholder," as there is no way to quantify benefits to the various affected publics over a reasonable time frame.

the subject from the administrator's point of view and from that of the prime consultant who has to apportion his budget.

Estimation of the environmental planning budget is commonly based on a percentage of the total planning budget. This can be misleading. For example, in a regional municipality where rapidly increasing population and urbanization called for meticulous highway, resources, and economic planning studies, costs were high. As a result, 3 to 4 percent of the total planning budget was sufficient to cover environmental planning costs. On the other hand, in a larger, less-populated part of Ontario where rural recreational land uses are dominant, 10 percent of the planning budget was allocated to environmental planning. This proved to be an inadequate allotment of funds because of the comparatively small amount of funds required for overall planning but the heavier time and travel costs involved in the larger area. The size of the area requiring coverage is more or less proportional to the cost of a study. Rather than using a fixed percentage of the planning budget, it would be better to apply a dollar figure per acre or hectare or per square mile or kilometer and adjust the budget as circumstances require.

For new town planning studies, subdivisions and institutions done at the ecosite and ecoelement scales (Fig. 3-1) and for highway feasibility studies, all in the conceptual planning stage, 12 percent is probably the usual percentage of the total study budget allocated to environmental issues. It varies between 8 percent and 15 percent depending on the particular area, the level of sophistication required, the issues identified, and the travel distance.

Thus, it is apparent that ecosystem processes must be brought into the economic and engineering paradigms, as these processes affect how humans use their natural, agricultural, and urban spaces. Work by systems engineers and ecologists offers promise in this regard (Koenig et al. 1975; Odum 1983), but interfacing is not easy.

Costs of Ecoplanning and Protection as a Proportion of Total Construction Costs

Figures combining environmental planning and environmental protection costs for a pipeline facility show that costs can be in the range of 7 percent of a multi-million-dollar construction project. A breakout of costs (Table 7-3) shows planning costs as 11 percent and protection costs as 89 percent of the total (Union Gas 1983). The most expensive items are wet-weather shutdown and topsoil stripping and replacement. Some costs, such as erosion control, are part of any normal construction cycle. That is, they are activities directly linked to operation of the facility, which would be done with or without environmental assessment and government inspection. Other items, such as environmental assessment and government

Table 7-3. Costs of ecoplanning and protection activities (1982) for 11 miles or 18 km of 42 inch or 1,067 mm gas pipeline construction (Union Gas 1983)[a].

Item	Cost
Environmental planning or ecoplanning costs	
Environmental study and environmental mapping	$ 69,000
Soil sampling and analysis	18,000
Subtotal	$ 87,000
Environmental protection costs	
Topsoil stripping	$ 82,000
Wet weather shutdown	324,000
Water sampling and analysis	2,000
Dust control	27,000
Excess subsoil removal	11,000
Stream crossing (estimated)	49,000
Topsoil replacement	82,000
Subsoil ripping and topsoil replacement	5,000
Stonepicking and trench redress	
the year following construction	71,000
Erosion control	8,000
Revegetation	13,000
Environmental monitoring and analysis	53,000
Subtotal	$727,000
Total	$814,000

[a] Costs compiled from invoices received for two sections of construction.

inspection. Other items, such as environmental monitoring, were added in the 1970s as part of energy board requirements.

Assuming from Table 7-3 a 100-foot or 30-meter right-of-way 11 miles or 18 kilometers long, the total area of the pipeline is 133 acres or 54 hectares. The expense of wet-weather shutdown to reduce soil compaction, topsoil stripping and replacement, subsoil ripping, and stone picking come to $4,240 per acre in 1983—all expenses related to restoring the resource to its original condition. Since much of the farmland along the line was worth between $1,000 and $2,000 an acre, alternative economic-environmental solutions would be to buy the right-of-way and convert it to permanent pasture, replant it and manage it as a prairie nature reserve, or offer farmers one-time cash settlement and let them do the renovation. If it is put in permanent pasture, it could be leased back to farmers. Most farmers, from my experience, would prefer to take the cash and do the work themselves. The pipeline companies would benefit as well if this approach were used, but the greater public interest of keeping farmland in top productive condition may not be accommodated. Alternative courses of action involve different value judgments but offer the promise of reducing costs overall.

Another way to look at the proportions of planning costs is by examining the total cost of a new home in Mississauga, Ontario. These figures (Table 7-4 show cost for both engineering and planning as less than 1 percent (actually 0.7 percent). The profit of 29.4 percent dwarfs these fees. Interestingly, the cost of the raw land is less (0.5 percent), demonstrating an interesting point: the land is essentially a "free" commodity; interest and various government levies and construction costs are where the action is.

Table 7-4. Typical 1975 cost figures for a new house in the Toronto area. (Reproduced with permission from the Central Mortgage and Housing Corporation, 1979.)

Components	Price	Percentage
Cash for land	$ 403	0.5
Mortgage	1,363	1.7
Mortgage interest	1,157	1.5
Servicing cost	5,230	6.7
Carrying costs—servicing	549	0.7
Survey fees	162	0.2
Engineering planning	520	0.7
Property taxes	632	0.8
Legal	300	0.4
Road/water levies	463	0.6
Lot levies	1,713	2.2
Transfer tax	98	0.1
Value increase (markup)	17,337	22.3
Mortgage interest	7,700	9.9
Property tax	50	0.1
Construction	31,850	40.8
Building permits	96	0.1
Transfer tax	363	0.5
Legal fees	782	1.0
Broker commission	1,560	2.0
Value increase (markup)	5,599	7.2
Total	77,927	100.0
Total markup	22,936	29.4

In another example of housing development in the one subdivision studied by MacNaughton (1973), engineering fees represented 3 percent of the total $198,500 in municipal services. Had professionally trained planners been hired and had environmental planning been done, using the costs published by Dorney (1977) for environmental planning, the total planning costs prorated for the 19.75 units would be $1,629, or 0.036 percent of the entire $445,666 package.* These total planning costs do not include any of the potential benefits that may have derived, for example, from the discovery of commercial sand-and-gravel deposits and from reduced storm drainage costs by using natural channels (see cost data in Table 7-2). Had an environmentally sensitive design been done, some of the differential profit between the Beechwood and the Lakeshore Village subdivisions might have been reduced (Table 7-2). The upper-income professional market all flowed to the Beechwood area, and the Lakeshore Village area was purchased by the lower end of the market.

In addition to the planning discussed above, if a complete environmental protection approach had been used in Lakeshore Village, the additional costs would be in the order of the environmental planning costs. These costs would be for environmental inspection, principally in the areas of erosion control and tree preservation. Furthermore, these costs would be invisible in the total budget.

The issue of costs for environmental planning and protection becomes one of perceived benefits by the developer and the city. If approvals were delayed by 6 days because the planning and engineering staff of the city asked relevant questions about environmental quality, the interest charges would more than equal all the environmental planning and environmental protection costs. This 6-day figure is very close to the 5-day figure developed earlier in the section "Costs and Savings— Conceptual Issues." Thus the conclusion that environmental planning should or should not be done may be viewed as a matter not of cost per se, but only of time. Minimal pressure from government can both encourage environmental management and make it cost-effective from the point of view of delays alone.

One other aspect of environmental protection costs is insurance costs associated with the facility operations mode. Environmental impairment coverage is expensive, especially if major spills and major litigation are probable. Avoiding or reducing such difficulties and associated costs requires environmental auditing procedures done by a third-party interdisciplinary group. Such groups might include lawyers, economists, chemical engineers, and environmental managers, the former providing the reductionist viewpoint, and the environmental manager and the lawyer the systems viewpoint. The recent Union Carbide disaster in

---

* To achieve these costs (in 1977 dollars), a minimum area of 100 acres is needed.

Bhopal demonstrates the difficulty and the risk in operating potentially dangerous facilities in urban areas.

Technical Information Agreement

The knowledge garnered in an ecosystem analysis can result, as I and others have demonstrated in many studies, in significant cost savings to the client. Unfortunately, the ecological practitioner or firm gets none of the economic advantages by so doing, unless the firm takes stock options in the development. This is a practice frowned on by some professionals for ethical reasons. If part of the savings were returned to the firm, a positive feedback loop would exist to develop sharper analytical skills regarding the identification of more savings in planning and construction costs.

Some technical equipment manufacturers, to assist in maintaining private sector technical personnel, sign a Technical Information Agreement (TIA) with clients so that as technical assistance is rendered, it is channeled back through the private firms. For the consulting environmental manager this concept might be a little difficult to negotiate with a client in a formal agreement. Informally there is a clear advantage to the client to continue to keep the environmental manager informed of key land use decisions and solicit assistance periodically until the project is complete, rather than only commissioning an ecological study as a "one-shot" consulting assignment.

If the client is apprised of the probable economic advantages in maintaining continuity between the environmental manager and the design/engineering/economic team, then a situation exists approaching the TIA concept used in manufacturing industries. In addition, a retainer could serve some of the same purposes as the TIA with the objective to provide a regionally based stable and skilled pool of personnel having an environmental systems capability.

Bibliography

Canadian National Energy Board (1984) Annual Report 1983. Ottawa: Canadian National Energy Board.

Dorney RS (1973) Role of ecologists as consultants in urban planning and design. Hum Ecol 1(3):183–200.

Dorney RS (1977) Biophysical and cultural-historic land classification and mapping for Canadian urban and urbanizing land. Proceedings, Workshop on Ecological Land Classification in Urban Areas, Canada Committee on Ecology Land Class. In: Land Class Series No. 3. Ottawa: Environment Canada, pp 57–71.

Dorney RS (1983) Costs of woodland, lawn restoration and maintenance compared (Ontario). Rest Mgmt Notes 1(4):22–23.

Dorney RS, George MG (eds) (1970) An Ecological Analysis of the Waterloo-South Wellington Region. Waterloo, Ont.: University of Waterloo, Division of Environmental Studies.

Grand River Implementation Committtee (1982) Summary and Recommendations. Grand River Basin Water Management Study. Cambridge, Ont. Grand River Cons Authority.

Horte VL (1983) Task Force Report on Pipeline Construction Costs. Ottawa: Energy, Mines and Resources Canada.

Koenig HE, Edens TC, Cooper WE (1975) Ecology, engineering and economics. Proc IEEE 63(3):501–511.

Loucks DR, Perkowski J, Bowie DB (1982) The impact of environmental assessment on energy project development. Downsview, Petro Canada (Calgary, Alberta) and York University.

MacNaughton I (1973) An economic and physical examination of urban open space with specific reference to natural floodplain parks. Unpublished MS thesis, University of Waterloo (Ontario), School of Urban and Regional Planning.

Ministry of Natural Resources (1981) Aggregate Resources Inventory of Pilkington Township, Wellington County, Southern Ontario. Queens Park, Ont.: Ontario Geological Survey, Aggregate Resources Inventory Paper 36.

Mutrie DF, Dorney RS (1981) Experience with environmental supervision of pipeline construction in Ontario. In: Environmental Concerns in Rights-of Way Management. Proceedings, Second Symposium. Palo Alto, CA: Electric Power Research Institute.

Nelson JG (1982) Public participation in comprehensive resource and environmental management. Sci Public Policy, 9(5):240–250.

Nelson JG, Day JC, Jessen S (1981) Regulation for environmental protection: The Nanticoke industrial complex (Ontario). Environ Mgmt 5(5):385–395.

Occidental Petroleum (1983) Annual Report. Los Angeles: Occidental Petroleum.

Odum HT (1983) Systems Ecology. New York: Wiley.

Ontario Treasury Board (1971) Review of Planning for the Grand River Watershed. Project No. 229. Queens Park, Ont.: Management Services Division.

Pushchak R, Burton I (1983) Risk and prior compensation in siting low-level nuclear waste facilities: Dealing with the NIMBY syndrome. Plan Can 23(3):68–79.

Schrecker TF (1984) Political Economy of Environmental Hazards. Study paper prepared for Law Reform Commission of Canada, Ottawa.

Thompson AR (1980) Environmental Regulation in Canada. Westwater Research Centre. Vancouver: University of British Columbia.

Union Gas (1983) One Year Post-Construction Monitoring Report of the Kerwood to Strathroy Gate Station and the Bright to Owen Sound Valve. Chatham, Ont.: Union Gas.

8
Professionalization: Reality and Prospects

The preceding chapters have presented a mixture of conceptual and practical issues to demonstrate the relevance of a new interdisciplinary, environmentally oriented profession. The critical elements posited can be summarized as follows:

- The increase in global and regional populations and their attendent industrial intensification have put stress on the structure and processes of the landscape. The ad hoc "remedies" so often used must be replaced with definitive and operational solutions.
- The science of ecology, combining an understanding of the bio-physical and human interactions, is well placed to understand land systems and to offer insightful commentary.
- The ethical human–nature interactions combine well with a series of technical principles to guide professional decisions on issues of the human and natural environment.
- The practice of environmental management breaks down into a series of "modes" principally: regional development, policy formulation, facility development, and facility operation. These modes include various planning sequences ranging from analysis, design, and approvals to various protection sequences beginning with contracting and construction and finishing with rehabilitation and monitoring.
- The organization of a professional practice in environmental management differs from that of the conventional practice of engineering and the design disciplines, as it requires developing an explicit

interdisciplinary, systems-oriented knowledge, and it requires the synergistic integration of many disparate technical and value-laden threads.

- The operational reality of an environmental management practice requires special conceptualization such as determining relevant scale, allowing for special seasonal requirements of fieldwork, and preparing reports that provide communication between experts and the public. However, it shares with engineering and the design professions considerations such as confidentiality, credibility, liability, and cash flow.

- The potential for significant economic savings in using environmental systems-oriented knowledge is demonstrated in my years of experience and that of others. Single discipline-driven development solutions can be wasteful by having overlooked often simple social, economic, or environmental dynamics and impacts. Thus, the normally perceived economic conflict between achieving environmental quality and economic development may disguise other socioeconomic or political issues. Skillfully integrated environmental design and assessment offer promise of reducing government costs and improving the general welfare at the same time.

Taken together, these critical elements suggest that a new profession—environmental management—can have wide benefits. This new profession is not intended to, nor could it attempt to, replace design disciplines (architects, landscape architects), planning, or engineering. The suggested format is one of professional cooperation, collaboration, and balance where ethical, procedural, design, and economic matters are evaluated in a human ecology paradigm, in contrast with information derived from using intuitive information and traditional reductionist approaches. The fitting together of such integrated information relevant to solving a new problem can result in dramatically improved solutions—a true synergy, satisfactory to all or many of the parties.

This new format, of course, readjusts both political and professional power. Agencies or development commissions dislike any challenge to their turf; professionals who have built reputations and sizable firms (and fortunes) based on a proven track record do not necessarily benefit by more creative solutions, nor do they welcome them. This means that institutional structures such as environmental assessment procedures, environmental hearing bodies, and environmental mediation guidelines are critical to the rebalancing of these well-entrenched, older power structures. Since democracies have pluralistic structures, such as opposition parties and public action groups, democratic societies are well placed to experiment with new institutional arrangements, arrangements that may be more appropriate to developing cost-effective and socially effective environmental solutions. Such solutions will not block change or development but rather require change to be justified and to be argued in holistic ways, often exposing the elements of power and assumed policy directions hidden from public view.

Governments, more comfortable with hierarchical distribution of power epitomized by strong central state planning structures, will find it useful to experiment deliberately with environmental management approaches to uncover what may be considerable inefficiencies embedded in present solutions. Such debate may, if well crafted, return considerable benefits by decreasing development costs, decreasing cleanup costs, negating retrofitting costs—the so-called externalities, commonly of a substantial and embarassing nature—and increasing the health and enjoyment of the people. Conceptually, these arguments are identified with four terms: appropriate technology, appropriate development, community development, and ecodevelopment—a kind of in-house jargon for environmental management.

In this general context, then, the final issues relevant to this new profession, evolving in many countries, are appropriate education and training, the size of markets for expertise, and development of professional associations. After pursuing these points I will finish by looking at the information-communication revolution and its substantial impact on global natural resources, their exploration, exploitation, and management, and on urban and regional landscape evolution.

Education and Training for the Environmental Management Professional

Based on almost two decades of graduate training of about 50 environmental professionals at the university level and on 15 years in environmental management consulting involving individuals from many discipline backgrounds, I believe the older-style or traditional education provided by undergraduate programs in agronomy, biology, civil engineering, forestry, geology, geography, landscape architecture, soil science, and wildlife management provides one reasonable option from which to begin career development in environmental mangement. The initial scientific bias is counterbalanced at the graduate level by explicitly exploring policy analysis, economics, and design.

The disciplines of economics, political science, and psychology, although having potential strengths in policy and economic areas, offer too little in the biophysical or field science area. For these students to make a successful transition into graduate studies in environmental planning and management, a postbachelor's makeup year or a graduate emphasis on courses in geomorphology, earth sciences, and the biological sciences provides a useful balance. For design professionals in landscape architecture, a similar balance between arts and science is requisite.

If some practical work experience in a resource management agency is added to an undergraduate's experience, either in science, arts, or landscape architecture, the student should be ready to undertake serious and complete graduate work in environmental mangement at the master's level. The question arising, then, is what kind of a graduate curriculum is likely to provide the balance of transdisciplinarity and systems perspectives, as proposed by Jantsche (1971) over a decade ago.

Since the experience discussed above demonstrates that environmental managers can come from many possible disciplines, a model master's curriculum is attempted (Table 8-1) for students whose undergraduate work was in science, arts, or design. This is then compared as a yardstick

Table 8-1. Model 2-year master's degree program in environmental management. Course weightings are shown as credit hours, assuming 30 credits to graduate.

Year 1[a]	
Professional practice of environmental management	6 credits
Environmental policy considerations	3
Environmental analysis—assessment (mediation, rehabilitation)	3
Workshop (collaborative group work)	3
Seminar (required, no credit)	NC
	15

Year 2	
Thesis (original research)	9
Environmental law	3
Elective, one of the following:	3
Environmental protection	
Resource economics	
Remote sensing	
Systems ecology	
Other, approved by thesis supervisor	—
	15

[a] An internship program 3 to 6 months in duration is strongly recommended between years 1 and 2, ideally oriented along the lines of the thesis research. Undergraduate work in ecological sciences or in economics and political science may also be required for those pursuing an interest in facility development and design (environmental assessment). For those interested in environmental protection work, such as waste management, a background in chemistry, toxicology, civil engineering, or hydrogeology would be beneficial. The thesis would further focus the professional interest of the graduate student.

with a number of existing programs in 1987, for illustrative purposes (Table 8-2). The actual programs selected are ones from well-known North American universities, known over many years for their sensitivity to resource management or to physical land use planning concerns. University-based institutes are not included, although it could be argued that superior work in interdisciplinary environmental studies is coming out of such groups. Their dependence on gurus and their reliance on home departments, which are sometimes lukewarm in support, make them an unstable academic structure, subject to rapid dissolution or evolution.

The point illustrated by these two tables is noteworthy. In Canada and the United States, a number of postsecondary educational institutions come close to meeting the general structure posited as suitable in Table 8-1. Only minor restructuring is required to offer a reasonably balanced program. Electives are available in other departments; work experience can, if available, provide a reasonable level of training without major department restructuring. Another interesting point is that the title of the program may belie the content, as the field of environmental management is only newly recognized.

The Market for Environmental Management Expertise

In examining employment opportunities, it is of course difficult to predict how many environmental professionals can be employed gainfully. Market forces, subject to cyclic changes, can produce restructuring of professional needs.

With professional membership in the Ontario Society for Environmental Management (OSEM) at about 60, and based on this group's knowledge of those in practice or teaching at universities who would be qualified to be members but are not members, the current 1987 total of environmental managers in Ontario stands at somewhere around 150. Thus, roughly the same as the Ontario Society for Landscape Architects (200) but considerably below the number in planning (862), architecture (1,600), and civil engineering (7,000).

For a field that began in 1965 to 1970 and blossomed during the environmental revolution of the 1970s, this rate of growth is substantial. The establishment of faculties of environmental studies and the passing of a plethora of environmental laws dealing with pollution abatement, impact assessment, and landscape rehabilitation have assisted the field to move rapidly forward. As consolidation begins, it is now an opportune time to examine old and new markets for expertise in environmental management.

A fair question to ask, then, is what rate of future growth, if any, can be expected for environmental managers? Certainly any answer is speculative, but if the expertise displayed is ethically driven, based on sound

Table 8-2. An evaluation of some current graduate programs and courses in planning and resources management, tested against the theoretical program and courses proposed in Table 8-1.[a]

University	Department or school[b]	Professional practice E.M.	Environmental policy	Environmental assessment	Environmental law	Environmental protection	Resource economics	Remote sensing	Systems ecology	Total score
Waterloo (Ontario, 1985–86)	Urban and Regional Planning	x	x	x	x			x		6
	Geography	x	x	x	x			x		6
	Civil Engineering				x	x			x	3
SUNY College of Environmental Science and Forestry (Syracuse 1982–83)	Environmental Science		x	x						2
	Environmental and Resource Engineering					x		x		2
University of Virginia (Charlottesville, 1987–88)	Environmental Sciences		x		x				x	3
University of Texas (Dallas, 1983–1985)	Community and Regional Planning			x	x					2

University	Department/Program									Score[a]
York (Toronto, Ontario, 1985–86)	Environmental Studies	x	x	x		x		x	x	5
North Carolina (Chapel Hill, 1983)	City and Regional Planning	x	x					x	x	3
	Environmental Sciences and Engineering	x	x	x	x		x	x	x	7
Johns Hopkins (Baltimore, 1983–84)	Geography and Environmental Engineering	x	x			x		x	x	4
Michigan (Ann Arbor, 1986–87)	Natural Resources	x	x	x		x		x	x	5

[a] Assuming a double wieght (course credit) for the course in professional practice, the highest possible score would be 9.

[b] Courses in calendar (date listed under university name) offered by the department or available within the faculty.

technical principles, and is economically beneficial, its acceptance by the public and private sectors should be good, and some growth can be expected. If, however, existing disciplines absorb its tenets and methodologies, environmental management could disappear or take on another form. Since the biological sciences, landscape architecture, civil engineering, and professional planning have not absorbed the field in the past 10 years, such a move on their part seems unlikely. Rather, the overlapping aspects of systems engineering, systems ecology, and environmental management suggest that a possible or partial convergence here is possible. In fact, a great deal of the environmental assessment work in the United States now is done by engineering-planning firms, not by independent environmental firms as in Canada. On the other hand, if environmental studies faculties, landscape architecture schools, and some natural resource–oriented schools develop a convergent thrust, their historic conceptual strength should allow them to maintain a significant share of the environmental management market for their graduates. When this will happen is uncertain; we are at a fork in the road.

Besides not offering suitable academic training, another reason why engineering sciences and the natural sciences may not be able to control the flow of expertise relates to the perception of corporate decision makers. Many of these corporate executives rightly perceive that environmental issues are not just technical problems for which engineers and scientists are trained to provide answers, but sociopolitical problems also. Such problems also require professionals who are skilled in dealing with the public, able to organize information suitable for hearings, and able to handle the news media. This means hiring professionals who understand where problems require technical answers and where problems require socioeconomic or culturally oriented or politically sensitive answers. Simply put, it is important to have both depth and breadth: one without the other makes it difficult, if not impossible, to operate in a complex decision-making world.

In Ontario, the development of a new professional organization, OSEM, was deliberately promoted, first to provide a *forum* where those professionally, as distinct from those academically, involved in environmental management could meet and discuss common issues. The existing professional organizations did not present this opportunity, as their memberships are unidisciplinary rather than transdisciplinary. The second reason for OSEM was to give public *identity* to the new profession. The Appendix has a description of OSEM membership and activities.

The Next 15 years—Planning the Unfinished Landscape

Given the transformation of industrial economics by the combination of the microchip and biotechnology, it is important to speculate on the effects such a transformation in information processing and genetic manipulation may have on the natural resources and landscape ecology of the globe. Some likely results are these:

- Increased need for a new definition of work and leisure resulting from increased automation of industrial production.
- Increased sophistication in locating and exploiting mineral and forestry resources and hence putting downside pressure on costs.
- Increased simplification of genetic diversity through breeding programs and loss of genetic diversity due to extinctions.
- Increased productivity of agricultural lands with attendant downward pressure on agricultural commodity prices.
- Increased efficiency in energy and material consumption in urban ecosystems due to the chip technologies.*

What this means is open to interpretation, but it could be dependent on what expectations we have about environmental quality concerns. For example, the uses of these new technologies can allow us to do more with less—joyous austerity (Dansereau 1973). Or theoretically, the poorer nations could more quickly develop national and regional economies by more efficient use of what are now scarce resources.

The negative side to this is that decisions affecting landscape processes can now be made more quickly as information moves more rapidly. Potential environmental impacts then could be both severe and large-scale unless some sophisticated talent is available for doing the critical environmental studies, asking the critical environmental questions, and designing effective environmental solutions. Societies (nations, states, regions) with environmental expertise to call on will be in a stronger position than those societies dependent on outside expertise. Because of the regional nature of ecosystem dynamics, such expertise must be locally trained and must have local experience to be effective.

Another issue that emerges from the effects of the microchip and biotechnology revolution is the likelihood of large-scale landscape rehabilitation. If calorie capture by crops per acre continues to rise as it has over the past four decades (see Dorney and Hoffman 1979), coupled with a drop in export demand (Insel 1985) resulting from the Green revolution, considerable agricultural land in North America (and elsewhere as well) can be and likely will be retired from production. The questions of what areas to retire, what vegetation to plant, what grazing pressure to allow, and what wildlife to manage on this new landscape will pose a unique and innovative opportunity for resource management agencies and their environmental managers.

To propose socially acceptable, environmentally acceptable, and economically satisfactory alternatives will require considerable vision to define the possible, the feasible, and the implementable. Instead of

---

* For example, reduction in waste of all kinds resulting from reuse, recycling, better precision tools, control of consumption, etc.

abandoning large and less productive land masses to the whims of natural succession, as happened, for example, in New England, Appalachia, and north central Ontario after World War II, environmental managers can take the present North American landscape mosaic— a product of the pressure for settlement and for the production of agricultural products— and mold this unfinished landscape into functional units of high ecological integrity. The necessary intensive zones for high agricultural and forestry productivity will be maintained. Here our knowledge of landscape ecology and biogeography will be tested empirically.

The quest for excellence—pursuing the planning and creation of this unfinished landscape—should be challenge enough for any group. The vision of a new profession dedicated to this challenging adventure is akin to the parallel adventure about to begin—the colonization of space. It is for this reason that I believe the next episode for resource management will not emphasize preservation and conservation, but instead will emphasize rehabilitation. So much of the world's surface is affected by humans that putting effort solely to preservation of unique landscapes is ultimately self-defeating; rehabilitation of leaky agroecosystems damaged by erosion or nutrient loss (Dorney 1984) and of urban areas such as waterfronts and waste management sites (Dorney 1985) is a chance for taking the inscape of the human mind and using it to create a new landscape, achieving thereby a harmony between human and nature in some kind of creative disequilibrium.

Bibliography

Dansereau P (1973) Inscape and Landscape. Massey Lectures. Ottawa: Canadian Broadcasting Corporation.

Dorney RS (1977) Planning for environmental quality in Canada: Perspectives for the future. Theme paper, Canadian Institute of Planners Annual General Meeting, Toronto.

Dorney RS (1984) Reclamation—sometimes the 'cure' is as bad as the 'disease.' Lands Arch 74(3):120.

Dorney RS (1985) Prospects for Urban Wildlife in the Year 2020. Proc of Wildlife Survivors in the Human Niche. Washington: Smithsonian—The National Zoo.

Dorney RS, Hoffman DW (1979) Development of landscape planning concepts and management strategies for an urbanizing agricultural region. Lands Plan 6:155–177.

Insel B (1985) A world awash in grain. Foreign Affairs 63(4):892–911.

Jantsche E (1971) Inter- and transdisciplinary university: A systems approach to education and innovation. Ekistics 193:430–437.

Afterword

Death took Robert Dorney before he completed the manuscript. The last chapter was somewhat less polished than those that go before, and he almost certainly would have wanted to elaborate on the ideas expressed in the last paragraph. On the other hand, I think we can safely assume that the paragraph introduces his final thoughts on the topic the manuscript addresses. The large philosophical idea with which the chapter concludes, alluding to the term "creative disequilibrium," could only have been discussed at the end of a work so thoroughly practical in approach and intent. This same practical character of the work strongly suggests, however, that Dorney would not have gone on at any great length in developing his final thoughts. Doing so would have unbalanced the manuscript.

No one can claim to know precisely what Dorney would have said in the few pages that remained to be written. Moreover, I think it would be a mistake to attempt to write them on his behalf. But beginning in the late 1960s and on and off over the years since, alas all too seldom, I belatedly realize, we discussed the sort of issues taken up in that last paragraph. It will perhaps not be amiss for me to suggest something of what he must have had in mind.

I shall focus on the two large ideas that he introduces. First, he claims that in the future, resource management must concentrate on rehabilitation rather than merely on preservation or conservation. Second, he describes the sort of rehabilitation he has in mind as directed toward achieving a special kind of harmony, marked by "creative disequilibrium." These are challenging and, perhaps to some, disturbing ideas.

But a little reflection should disclose that they are also powerful and altogether positive ideas. I will say a word or two about each separately.

Preservation and conservation, he notes, are ultimately self-defeating, if we attempt to do nothing more than preserve and conserve. This is so because the major forces at work in our world press against the areas we would preserve and conserve. Accordingly, the practical effect of focusing exclusively on preservation and conservation must be similar to that of attempting to hold back the tide. The effect can only be to slow it, not to arrest or much less reverse it. So, in that sense, preservation and conservation, if they are the resource manager's sole strategy, are self-defeating.

A similar point is made by Barry Commoner. In a speech titled "The Failure of the Environmental Effort," delivered to the staff of the U.S. Environmental Protection Agency in January 1988, Commoner contrasts the uses made by the EPA of the concept of "acceptable risk" with the preventive approach to public health. The latter approach works toward a positive goal, and success is measured by the rate at which it is achieved. To apply the preventive approach of public health to environmental problems would be to strive for a "continuous improvement in environmental quality." When standards of acceptable risk form the regulatory norms, effort is expended exclusively on minimizing the damage rather than on a continuous advance toward a "healthy" environment.

As norms, preservation and conservation work in the way that acceptable risk does. Since not everything can be preserved, the best one can achieve is to minimize the damage; no gains can be made. To focus resource management on rehabilitation, by contrast, is to give it the same positive and forward-looking orientation that typifies the preventive approach to public health.

The second large idea is that of creative disequilibrium. It would be a mistake to conclude that, because "disequilibrium" is the last word in the manuscript, it closes on a negative or pessimistic note. The disequilibrium is to be creative. That changes everything. But what can be meant by this unexpected reference to disequilibrium as something to be sought after? The clue, I think, is provided by the preceding words: "taking the inscape of the human mind and using it to create a new landscape, achieving thereby a harmony between human and nature. . . ."

Dorney's underlying concept appears to be that the resource manager and humans in general are not to situate themselves as outside the environments they manage, responsible only for keeping those environments in (or returning them to) their original, pristine condition. Rather, we are part of the scene and bring to it what is distinctively ours— creativity. The objective of resource management must be to create a whole that contains, respects, and expresses the presence of humans.

So a relationship of humans with the rest of nature, considerably more complex than the two as usually emphasized, is contemplated. Usually, the exploiters are opposed to the custodians: at one end of the scale, those who see nature as a warehouse of resources to be used up in pursuit of human ends; at the other end, those who would fence off the nonhuman parts of nature and demand of us nothing more than the role of the good steward.

Creative disequilibrium establishes a kind of harmony, and it relates humans with the rest of nature. The stress on harmony expresses Dorney's rejection of the exploitative view of our relationship with nature. The stress on disequilibrium expresses his rejection of the custodial view of that relationship.

There is an obvious connection between the two large ideas under discussion here. In condemning exclusive reliance on preservation and conservation, Dorney is already rejecting the custodial view. By advocating rehabilitation, he is asserting the importance of nonexploitative human intervention. "Creative disequilibrium" is then his term for the character of such intervention.

On Dorney's account, environmental management is the practice that aims at creative disequilibrium. The environmental manager is neither an exploiter nor a custodian. If we take seriously the reference to using the "inscape" of the human mind to create a new landscape, then we will say that Dorney envisaged the environmental manager as not merely a professional, but an artist.

Lawrence Haworth, Professor of Philosophy and cross-appointed in the School of Urban and Regional Planning at the University of Waterloo, was a close colleague of Robert Dorney. Haworth has published *The Good City* (1963, Indiana University Press), *Decadence and Objectivity* (1977, University of Toronto Press), and *Autonomy* (1986, Yale University Press).

Appendix

Developing a Professional Association—The Ontario Society for Environmental Management

Since few environmental professional associations have been attempted, it may be useful to briefly review the experience in Ontario with the Ontario Society for Environmental Management (OSEM). In the mid-1970s, the National Association for Environmental Professionals (NAEP) was formed in the United States. As it is the only national association, to my knowledge, in existence, it too could be a useful prototype for examination. Similarly, some Australians at the University of Brisbane (Westoby 1984) are interested in organizing such a professional association and may have done so by now. Thus there are a few models in operation and some under discussion, but none with more than 10 years' experience.

The organization of OSEM, a provincial-level organization, took place through a series of workshops held between academics, consultants, and key government personnel in 1975 and 1976. As a new provincial environmental impact assessment act was being prepared, it was apparent that upwards of $5 million would be thrust into the consulting markets, markets essentially unorganized to receive such an infusion of capital. To maintain public credibility, some professional structure was considered desirable.

A steering committee prepared a constitution in which is imbedded a certification process. The certification process parallels that of the Canadian Institute of Planners: a full member must have a bachelor's degree and 4 years' professional experience or master's degree and 2 years' professional experience. A doctorate, because of its specialized nature, was considered equivalent to a master's degree. A category was also created for provisional and student members.

Total membership 10 years later has been maintained at about 65, with 20 to 25 student members. The 1981 to 1983 recession saw considerable professional retrenchment as construction and development died, but the membership did not dip. Although a small group, it is a reasonably loyal one.

Since Ontario is a large physical space, the group's core is based in Toronto and Waterloo. This makes it difficult for members in outlying areas, such as Ottawa and Kingston, to make meetings on a regular basis. Dues, currently at $45 per year, provide too few funds to support a journal. Such a level of funding defrays costs for four to six meetings held from fall to spring. In addition, a small secretariat is maintained with a part-time administrative assistant.

The effectiveness of OSEM, given its small size, is hard to measure. However, by preparing briefs for government, meeting with ministers on occasion, and testifying before hearings and commissions, our collective position is heard. It seems probable that without OSEM the public could have been inundated, beyond what it was, with environmentalists with little or no training or experience. Of course such impressions are anecdotal and hard to measure indeed. Certainly the key to any organization is strong, effective leadership. Any structure, no matter how well conceived, can founder in its formative stage. The issue of being self-serving and not acting in the public interest has been raised at many meetings and in other forums as well (Westoby 1984). As long as the technical body of knowledge on which environmental mangement rests is not ideologically restricted by having a professional body in place, as happened apparently in medicine (Westoby 1984), the public can benefit clearly by such a professional body.

Bibliography

Westoby M (1984) Some dangers for ecology on the way to becoming a profession. Austral J Ecol 9:301–308.

Index